Analytical Methods for Food and Environmental Pollutants: Current and Future Perspectives

Analytical Methods for Food and Environmental Pollutants: Current and Future Perspectives

Guest Editors

Julia Martín
Esteban Alonso

MDPI

Basel • Beijing • Wuhan • Barcelona • Belgrade • Novi Sad • Cluj • Manchester

Guest Editors
Julia Martín
Departamento de Química Analítica
Universidad de Sevilla
Seville
Spain

Esteban Alonso
Departamento de Química Analítica
Universidad de Sevilla
Sevilla
Spain

Editorial Office
MDPI AG
Grosspeteranlage 5
4052 Basel, Switzerland

This is a reprint of the Special Issue, published open access by the journal *Molecules* (ISSN 1420-3049), freely accessible at: https://www.mdpi.com/journal/molecules/special_issues/I0225DF8X0.

For citation purposes, cite each article independently as indicated on the article page online and as indicated below:

Lastname, A.A.; Lastname, B.B. Article Title. *Journal Name* **Year**, *Volume Number*, Page Range.

ISBN 978-3-7258-7004-2 (Hbk)
ISBN 978-3-7258-7005-9 (PDF)
https://doi.org/10.3390/books978-3-7258-7005-9

Cover image courtesy of Julia Martín and Esteban Alonso

Contents

About the Editors . . . . **vii**

Preface . . . . **ix**

Stephanie D. Nauth and Andres D. Campiglia
Sorption of Polycyclic Aromatic Sulfur Heterocycles (PASH) on Nylon Microplastics at Environmentally Relevant Concentrations
Reprinted from: *Molecules* **2024**, *29*, 1653, https://doi.org/10.3390/molecules29071653 . . . . . . **1**

Carmen Mejías, Marina Arenas, Julia Martín, Juan Luis Santos, Irene Aparicio and Esteban Alonso
Multiclass Analysis for the Determination of Pharmaceuticals and Their Main Metabolites in Leafy and Root Vegetables
Reprinted from: *Molecules* **2024**, *29*, 3471, https://doi.org/10.3390/molecules29153471 . . . . . . **10**

Michelle Peter and Christoph Müller
Systematic Comparison of Extract Clean-Up with Currently Used Sorbents for Dispersive Solid-Phase Extraction
Reprinted from: *Molecules* **2024**, *29*, 4656, https://doi.org/10.3390/molecules29194656 . . . . . . **21**

Xinyuan Bi, Chi Zhang, Xiaorong Xue, Shangjun Su, Zhiping Yang, Xu Jing and Qiang Zhang
Emulsive Liquid–Liquid Microextraction for the Determination of Phthalic Acid Esters in Environmental Water Samples
Reprinted from: *Molecules* **2024**, *29*, 5908, https://doi.org/10.3390/molecules29245908 . . . . . . **39**

George T. Knecht, Stephanie D. Nauth, Juan C. Gomez Alvarado, Anthony M. Santana, Hector C. Goicoechea and Andres D. Campiglia
Unambiguous Determination of Benzo[a]pyrene and Dibenzo[a,l]pyrene in HPLC Fractions via Room-Temperature Fluorescence Excitation–Emission Matrices
Reprinted from: *Molecules* **2025**, *30*, 1550, https://doi.org/10.3390/molecules30071550 . . . . . . **50**

Tamara Pócsová, Senad Okanovič and Svetlana Hrouzková
Green Analytical Method Using Single-Drop Microextraction Followed by Gas Chromatography for Nitro Compound Detection in Environmental Water and Forensic Rinse Water
Reprinted from: *Molecules* **2025**, *30*, 1894, https://doi.org/10.3390/molecules30091894 . . . . . . **61**

Michaela Zeiner, Viktor Sjöberg and Helena Olsman
Metal Uptake by Birches and Scots Pines Grown on a Porcelain Landfill
Reprinted from: *Molecules* **2025**, *30*, 2196, https://doi.org/10.3390/molecules30102196 . . . . . . **77**

Phannika Tongchai, Nootchakarn Sawarng, Anurak Wongta, Udomsap Jaitham, Kunrunya Sutan, Saweang Kawichai, et al.
Optimization and Validation of a QuEChERS-Based Method Combined with Gas Chromatography–Tandem Mass Spectrometry for Analyzing Pesticide Residues in Edible Insect Samples
Reprinted from: *Molecules* **2025**, *30*, 2293, https://doi.org/10.3390/molecules30112293 . . . . . . **92**

Natalia Manousi, Apostolia Tsiasioti, Abuzar Kabir and Erwin Rosenberg
Determination of Phenylurea Herbicides in Water Samples by Magnet-Integrated Fabric Phase Sorptive Extraction Combined with High Performance Liquid Chromatography
Reprinted from: *Molecules* **2025**, *30*, 3135, https://doi.org/10.3390/molecules30153135 . . . . . . **107**

About the Editors

Julia Martín

Julia Martín is a Full Professor of Analytical Chemistry at the University of Seville. Her research focuses on innovative analytical methods and the environmental behavior of emerging contaminants, including their detection, risk assessment, and removal. She has carried out research stays at leading international institutions, collaborated extensively with industry and public agencies, and participated in 25 funded projects. She is the author of over 150 scientific papers and 26 book chapters, and appears in Stanford University's World's Top 2% Scientists List.

Esteban Alonso

Esteban Alonso is a Full Professor of Analytical Chemistry and Head of the Department of Analytical Chemistry at the University of Seville. He leads the Industrial and Environmental Chemical Analysis Research Group (ANQUIMED), serves as a Member of the University of Seville Governing Council, and represents the University of Seville in the Andalusian Regional Group of the Spanish Society of Analytical Chemistry.

Preface

This Reprint presents a curated selection of recent advances in analytical chemistry focused on the detection, characterization, and interpretation of pollutants in food and environmental systems. Its scope reflects the growing scientific and societal need for robust analytical data as the variety, complexity, and ubiquity of contaminants continue to expand, often at trace and ultratrace concentrations. The aim of this Reprint is to provide a unified collection of studies that demonstrate how innovative sampling strategies, refined sample preparation methodologies, advanced analytical instrumentation, and powerful chemometric tools collectively strengthen our capacity to monitor and understand pollutant behavior.

The motivation for assembling this Reprint stems from the rapid emergence of new contaminants and the ongoing evolution of analytical techniques designed to address them. With contributions that span conceptual developments, methodological refinements, and applied research, the Reprint offers readers a comprehensive and integrated view of the current state of the field.

This Reprint is intended for researchers, laboratory professionals, regulatory authorities, educators, and students who seek a deeper understanding of contemporary approaches to pollutant analysis. By highlighting methodologies that enhance accuracy, reliability, and interpretative value, it aims to support scientific progress and to contribute to the protection of food quality and environmental integrity.

Julia Martín and Esteban Alonso
Guest Editors

molecules

MDPI

Article

Sorption of Polycyclic Aromatic Sulfur Heterocycles (PASH) on Nylon Microplastics at Environmentally Relevant Concentrations

Stephanie D. Nauth and Andres D. Campiglia *

Department of Chemistry, University of Central Florida, Physical Sciences Building. 4111, Orlando, FL 32816, USA; st655863@ucf.edu
* Correspondence: andres.campiglia@ucf.edu

Abstract: Microplastics have garnered an infamous reputation as a sorbate for many concerning environmental pollutants and as a delivery vehicle for the aquatic food chain through the ingestion of these contaminated small particulates. While sorption mechanisms have been extensively studied for polycyclic aromatic hydrocarbons, polycyclic aromatic sulfur heterocycles (PASHs) have not been investigated, partly due to their low concentrations in aquatic ecosystems. Herein, an analytical methodology is presented for the analysis of dibenzothiophene, benzo[b]naphtho[1,2-b]thiophene, benzo[b]naphtho[2,1-b]thiophene, benzo[b]naphtho[2,3-b]thiophene, chryseno[4,5-bcd]thiophene and dinaphtho[1,2-b:1′,2′-d]thiophene at relevant environmental concentrations based on solid phase extraction and high-performance liquid chromatography. The sorption uptake behavior and the sorption kinetics of the three benzo[b]napthothiophene isomers were then investigated on nylon microplastics to provide original information on their environmental fate and avoid human contamination through the food chain. The obtained information might also prove relevant to the development of successful remediation approaches for aquatic ecosystems.

Keywords: polycyclic aromatic sulfur heterocycles; sorption; sorption uptake; kinetics; nylon microplastics

1. Introduction

Microplastics have been a concern in aquatic environments throughout the years. With dimensions ranging from 0.1 to 5000 μm, the small size of these particulates makes their removal from environmental waters challenging [1] and facilitates their ingestion by aquatic organisms [2]. The rather large surface area of microplastics promotes the uptake of toxic contaminants from aquatic environments and enhances their role as chemical vectors for food chain contamination [3]. For instance, specific surface areas for polyamide microplastics have been reported in the range of 0.017–0.544 m^2/g for particles sizes varying from 4000 to 29.7 μm, respectively [4]. Organic pollutants known to sorb onto microplastics include polycyclic aromatic hydrocarbons (PAHs) [3], pharmaceuticals [5], polychlorinated biphenyl [6] and polyhalogenated carbazole [7]. Their sorption onto microplastics occurs due to several retention mechanisms, which include a combination of pore filling, hydrogen bonding, pi-pi interactions, electrostatic interactions and van der Waals forces [8].

The presented studies deal with the sorption of polycyclic aromatic sulfur heterocycles (PASHs) onto nylon microplastics. Similar to their parent homocyclic species (PAHs), which contain only carbon and hydrogen, PASHs are usually formed during the incomplete combustion or pyrolysis of organic matter. In addition to their mutagenic and carcinogenic properties, some of them appear on pollutant lists around the world [9–11]. Specifically relevant to food chain contamination is the exacerbated ability of PASHs to bioaccumulate in aquatic organisms [12], which results from the presence of the sulfur atom in their heterocyclic molecular structure.

According to our literature search, this is the first report on the sorption of PASHs onto microplastics. To achieve low concentrations of PASHs in aqueous environments [13,14], a method was developed based on solid-phase extraction (SPE) and high-performance liquid chromatography (HPLC) with fluorescence detection. The pre-concentration factors achieved with the SPE procedure, and the appropriate choice of excitation and emission wavelengths made the analysis of several PASHs at parts per trillion (ppt) concentration levels possible. These included dibenzothiophene (DBT), benzo[b]naphtho[1,2-b]thiophene (BbN12T), benzo[b]naphtho[2,1-b]thiophene (BbN21T), benzo[b]naphtho[2,3-b]thiophene (Bb23T) chryseno[4,5-bcd]thiophene (C45T) and dinaphtho[1,2-b:1′,2′-d]thiophene (DiN1212T).

The SPE-HPLC method was then applied to study the sorption uptake and sorption kinetics of BbN12T, BbN23T and BbN21T onto nylon 11 and nylon 6,12. Nylon or polyamide is a popular type of plastic used in fishing gear for both commercial and recreational purposes. Since 18% of marine plastic waste is due to the fishing industry [15], the present studies provide relevant information regarding the contamination issue of microplastics in aquatic ecosystems.

2. Results and Discussion

2.1. HPLC Analytical Figures of Merit

HPLC conditions were optimized for the separation and detection of the six PASHs studied in this investigation. Figure S1 shows a typical chromatogram obtained from a synthetic mixture with the separation conditions outlined in Section 3.3. Since complete separation was obtained in all cases, these chromatographic conditions were kept constant for the remainder of this study.

The excitation and emission wavelengths for fluorescence detection were selected from signal-to-noise (S/N) ratios measured at the maximum intensities of chromatographic peaks (S) and the baseline noise (N) collected for 0.5 min prior to each chromatographic peak in the chromatograms. Table S1 summarizes the excitation and emission wavelengths that provided the highest S/N ratio for each PASH, along with their time window for chromatographic elution.

The analytical figures of merit for HPLC analysis are summarized in Table 1. No attempts were made to experimentally determine the upper concentration limit of the linear dynamic range. The correlation coefficients that are close to unity confirm the linear dynamic ranges (LDRs) with at least two orders of magnitude and the relative standard deviations (RSDs) demonstrate the good reproducibility of measurements at the parts-per-billion (ng/mL) concentration levels. The limits of detection (LODs) and the limits of quantitation (LOQs) were below 1 and 3 ng/mL, respectively. These LODs and LOQs are well above the parts-per-trillion concentration levels of PASHs in aquatic ecosystems [13,14].

Table 1. HPLC analytical figures of merit.

PASH	Retention Time [1] (min)	LOD [2] (ng/mL)	LOQ [3] (ng/mL)	LDR [4] (ng/mL)	% RSD [5]
DBT	3.87 ± 0.022	0.9	3.0	3.0–200	5.64
BbN12T	5.68 ± 0.078	0.3	0.9	0.9–200	3.53
BbN23T	7.07 ± 0.135	0.2	0.8	0.8–200	0.30
BbN21T	8.78 ± 0.196	0.3	0.9	0.9–200	1.94
C45T	13.24 ± 0.346	0.5	1.8	1.8–200	4.10
DiN1212T	26.33 ± 0.868	0.7	2.2	2.2–200	3.75

[1] Retention times are the average of three sample injections; sample volume = 20 μL. [2] Limit of detection (LOD) was calculated using LOD = $3S_B/m$, where S_B is the standard deviation of the blank using the following formula: 1/5 ($Noise_{max} - Noise_{min}$). The noise was recorded for 0.5 min at the base peak of each chromatographic peak. [3] Limit of quantitation (LOQ) was calculated using LOQ = $10S_B/m$. [4] Linear Dynamic Range (LDR), in ng/mL, extends from the LOQ to an arbitrarily chosen upper linear concentration. [5] Relative Standard Deviation (RSD) = $S/I \times 100$, where I is the average intensity and S is the standard deviation of the intensity calculated from three measurements at the middle concentration.

2.2. HPLC-SPE Analytical Figures of Merit

Figure S2 provides a schematic diagram of the SPE procedure coupled to HPLC to achieve environmental concentrations of PASHs. The use of heptanol prior to sample pre-concentration removed unknown fluorescence impurities present in the SPE cartridges that co-eluted with several PASHs.

Figure S3 shows a typical SPE-HPLC chromatogram obtained from the pre-concentrated sample. Table 2 summarizes the SPE-HPLC AFOMs for the studied compounds. As expected, the retentions times were consistent with those in Table 1 and the LODs and LOQ were improved by at least one order of magnitude. The LDRs of the calibration curves and the acceptable reproducibility of measurements (RSD at medium linear concentrations demonstrate the ability to perform an SPE-HPLC analysis of PASHs at environmentally relevant concentration levels). Although BbN12T was the only PASH to present an analytical recovery statistically equivalent to 100% (t_{exp} = 1.98, t_{crit} = 4.30, P = 95%; N = 3), the low standard deviations of all the analytical recoveries show the satisfactory precision of measurements for pursuing sorption studies at environmentally relevant concentrations.

Table 2. SPE−HPLC analytical figures of merit.

PASH	Retention Time [1] (min)	LOD [2] (pg/mL)	LOQ [3] (pg/mL)	LDR [4] (pg/mL)	RSD [5] (%)	Analytical Recovery [6] (%)
DBT	3.73± 0.015	0.05	0.2	0.2–75	5.0	90.5 ± 3.1
BbN12T	5.57 ± 0.120	3	10	10–75	1.1	101.6 ± 1.4
BbN23T	7.02 ± 0.125	0.7	2	2–100	3.0	88.2 ± 3.4
BbN21T	8.72 ± 0.159	0.6	2	2–100	3.9	77.0 ± 2.7
C45T	13.11 ±0.405	2	8	8–75	7.4	68.5 ± 4.1
DiN1212T	25.89 ± 0.674	2	7	7–75	1.1	69.9 ± 2.5

[1] Retention Times are the average of three sample injections; sample volume = 20 μL. [2] Limit of detection (LOD) was calculated using LOD = $3S_B/m$, where S_B is the standard deviation of the blank using the following formula: $\frac{1}{5}(Noise_{max} - Noise_{min})$. The noise was recorded for 0.5 min at the base peak of each chromatographic peak. [3] Limit of quantitation (LOQ) was calculated using LOQ = $10S_B/m$. [4] Linear Dynamic Range (LDR) in ng/mL; extends from the LOQ to an arbitrarily chosen upper linear concentration. [5] Relative Standard Deviation (RSD) = S/I × 100, where I is the average intensity and S is the standard deviation of the intensity calculated from three measurements at the middle concentration. [6] Analytical recoveries (AR) were calculated with the formula AR = $C_S/C_M \times 100$, where Cs is a medium linear concentration of the standard submitted to the entire experimental procedure and C_M is the concentration obtained with the method. The reported values are the averages of three experimental runs.

2.3. *Sorption Uptake by Nylon Microplastics*

BbN12T, BbN23T and BbN21T were selected for these studies. These three PASHs have the same molecular weight (234 g·mol^{-1}) and the same water solubility (2.291 mmol/L) [16–18]. Their molecular lengths (L) and thicknesses (T) varied as follows: BbN12T (L = 12.72 Å and T = 4.39 Å), BbN21T (L = 13.66 Å and T = 4.06 Å) and BbN23T (L = 13.90 Å and T = 4.06 Å) [19,20].

Aqueous solutions containing the three isomers and microplastic particles were thoroughly shaken for various exposure times and the isomer concentrations were monitored via SPE-HPLC. The sorption uptake (q_p) of each PASH was calculated with the following formula [21]:

$$q_p = \frac{(c_t - c_{aq}) \times 0.100\ L}{m_p} \quad (1)$$

where q_p represents the mass of PASH adsorbed on the surface of the pellet (ng/g), c_t is the initial concentration of each PASH in aqueous solution (ng/L), c_{aq} is the PASH concentration in the aqueous solution after a certain exposure time (ng/L) and m_p is the mass of the microplastics (g) used for the experiments.

The maximum sorption uptake for the tested plastics is shown in Figure 1. The maximum sorption uptake was determined using Equation (1) and the standard deviations were calculated using the values from Table S2 and the Equation (S1) through Equation (S3). The figure shows that Nylon 6,12 shows a higher mass sorbed for all the PASH isomers. BbN23T showed the highest sorption between both plastics while BbN21T showed the least sorption. These trends indicate that the type of nylon impacts the sorption uptake of these pollutants, and that the pollutant structure also dictates the sorption onto microplastics. In addition, Table S3 shows the masses for the maximum sorption uptake. The masses sorbed for BbN23T and BbN12T are statistically different from each other, while the masses sorbed for BbN21T for both Nylon 11 and Nylon 6,12 are statistically equivalent.

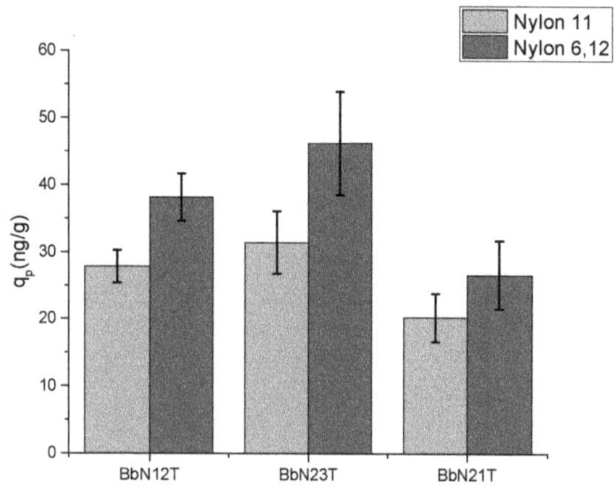

Figure 1. Maximum sorption uptake for BbN12T, BbN23T and BbN21T onto Nylon 11 and Nylon 6,12.

Figure 2 shows the time needed to achieve the maximum sorption. For all PASHs, Nylon 6,12 was shown to be faster than Nylon 11. With Nylon 6,12, the sorption the maximum sorption occurred in less than 2 h for all three isomers. It took almost more than double the time for Nylon 11 to reach the maximum sorption uptake compared to Nylon 6,12.

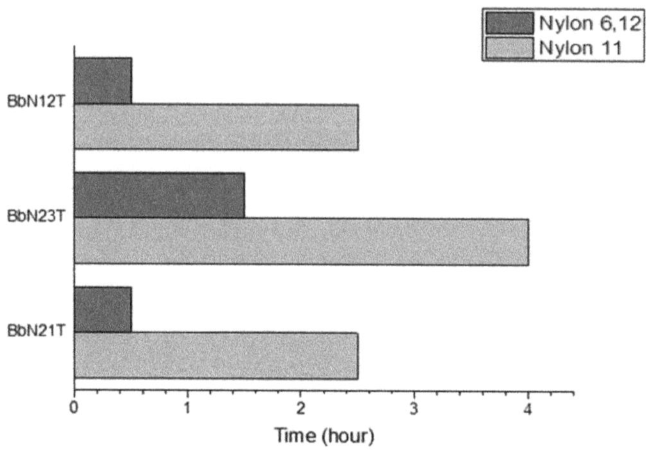

Figure 2. Time taken for maximum sorption to occur of BbN12T, BbN23T and BbN21T onto Nylon 11 and Nylon 6,12.

2.4. Sorption Kinetic Modeling

The sorption of the PASHs onto the microplastics was further analyzed using pseudo-first-order and pseudo-second-order fittings. Both fittings are commonly used to understand the mechanisms driving the sorption. The sorption uptakes established an equilibrium that was further proven by the statistical equivalence of the last two time points.

When the sorption behavior follows pseudo-first-order, adsorption occurs through diffusion [21]. Pseudo-first-order is represented by Equation (2).

$$\log\left(q_e - q_p\right) = \log q_{eq,1} - \frac{k_1}{2.303}t \tag{2}$$

where q_e = mass on the pellet at equilibria (ng/g), q_p = mass on pellet after exposure time (ng/g), $q_{eq,1}$ =mass on the pellet at equilibria, determined by the pseudo-first-order model, and k_1 = rate constant for the pseudo-first-order model. Figures 3 and 4 show the fittings for both Nylon 11 and Nylon 6,12. None of the isomers show any linearity with the model; therefore, it can be concluded that the absorption mechanism is not diffusion-based.

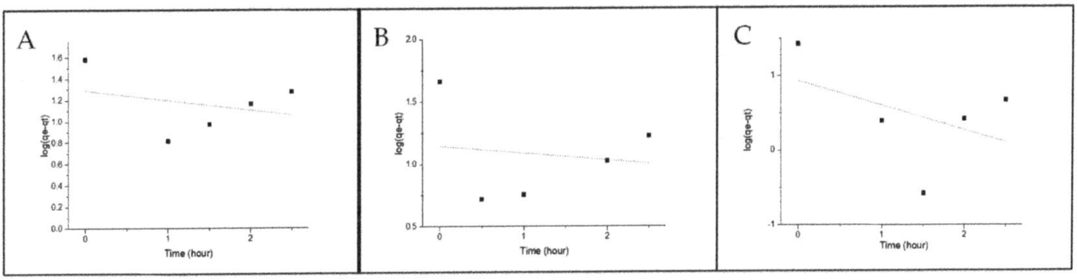

Figure 3. Pseudo-first-order fitting onto Nylon 6,12 for (**A**) BbN12T, (**B**) BbN23T and (**C**) BbN21T.

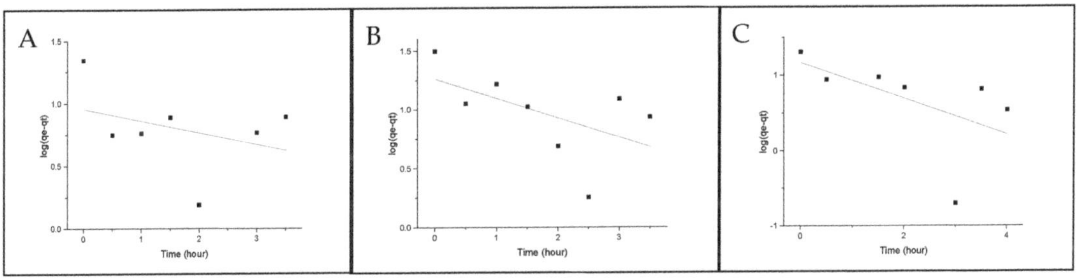

Figure 4. Pseudo-first-order fitting onto Nylon 11 for (**A**) BbN12T, (**B**) BbN23T and (**C**) BbN21T.

Pseudo-second-order kinetic modeling shows that the absorption rate is dependent on the capacity for sorption and not the concentration of the absorbate [21]. This model is represented by the following equation:

$$\frac{t}{q_t} = \frac{1}{k_2 q_{e,model}^2} + \frac{1}{q_{e,model}}t \tag{3}$$

where $q_{e,\,model}$ is the sorption capacity at equilibrium, q_t is the amount of the analyte sorbed at each time point, k_2 is the pseudo-second-order rate constant and k_2q_e is the initial sorption rate. Figures 5 and 6 show the fitting for both Nylon 6,12 and Nylon 11. All the isomers show great linearity, which indicates good agreement with the model.

Table 3 summarizes the rate constants and the initial sorption rates. Nylon 6,12 showed a higher initial sorption rate than Nylon 11. The model predicts that BbN23T will have the highest sorption capacity at equilibrium, which is confirmed by the sorption uptake showing that BbN23T has the largest uptake on the microplastics.

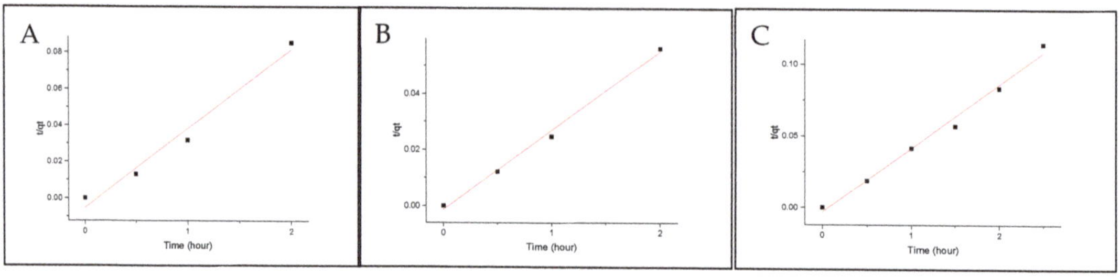

Figure 5. Pseudo Second Order fitting onto Nylon 6,12 for (**A**) BbN12T, (**B**) BbN23T, and (**C**) BbN21T.

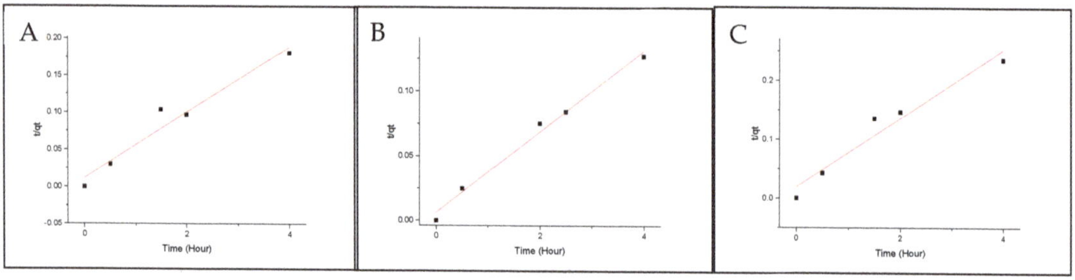

Figure 6. Pseudo-second-order fitting onto Nylon 11 for (**A**) BbN12T, (**B**) BbN23T and (**C**) BbN21T.

Table 3. Pseudo-second-order fitting parameters.

Type of Microplastics	PASHs	R^2 [a]	$q_{e,model}$ [b] (ng/g)	$k_2q^2_{e,model}$ [c] (ng/g) × h	k_2 [d] (g/ng) × h
Nylon 6,12	BbN12T	0.9778	23.87	181.8	0.3191
	BbN23T	0.9953	35.46	714.3	0.5681
	BbN21T	0.9949	24.39	1111	1.8676
Nylon 11	BbN12T	0.9975	22.52	217.4	0.4287
	BbN23T	0.9895	31.95	163.9	0.1606
	BbN21T	0.9515	17.21	51.81	0.1749

[a] R^2 = Correlation coefficient obtained by fitting data to the pseudo-second-order model. [b] $q_{e,model}$ = PASHs-sorbed amount at equilibrium obtained from the pseudo-second-order model fitting using Equation (2). [c] $k_2q^2_{e,model}$ = initial sorption rate, obtained from the intercept of Equation (2). [d] k_2 = Rate constant of the pseudo-second-order model, obtained from the intercept of pseudo second-order model's Equation (2).

3. Materials and Methods

3.1. Materials and Chemicals

DBT and BbN12T (purity = 99%) were purchased from Sigma Aldrich (Milwaukee, WI, USA). BbN23T (purity > 99%) and BbN12T (purity > 99%) were acquired from the European Commission (Brussels, Belgium). C45T and DiN1212T were both obtained from the National Institute Standards and Technology (NIST) (Gaithersburg, MD, USA) and used as-received. The molecular structures for these compounds are presented in Figure S4.

Nylon 11 and Nylon 6–12 were obtained from Sigma Aldrich as pellets, with particle sizes of 3 mm and 2 mm, respectively. The molecular structures for these compounds are shown in Supplemantal Information Figure S5. Physicochemical properties for both nylons are displayed in Table S4 [22–24]. The averaged dimensions of the microplastic pellets used are shown in Table S5.

HPLC-grade n-hexane, n-heptane, n-octane, n-nonane and methanol were acquired from Sigma Aldrich. HPLC-grade heptanol was purchased from Acros Organic (Geel, Belgium). HPLC-grade acetonitrile (ACN) was obtained from VWR (Radnor, PA, USA). Nanopure water obtained from a Barnstead Nanopure Infinity (Dubuque, IA, USA) water system was used throughout all the experiments. Waters (Milford, MA, USA) Sep-Pak Plus C-18 cartridges were used on a vacuum manifold.

3.2. Stock Preparation

Stock solutions of PASHs were prepared with the following solvents: n-heptane (DBT) n-octane (BbN12T, BbN21T, C45T and DiN1212T) and n-nonane (BbN23T). All stock solutions were kept in the dark at 4 °C. Working solutions of pure PASHs were prepared daily via evaporation of the n-alkane solvent under a gentle stream of nitrogen gas (100% purity). In all cases, the reconstituting solvent was ACN.

3.3. HPLC Analysis

All PASHs concentrations were monitored via high-performance liquid chromatography (HPLC) with an HPLC system from Hitachi (San Jose, CA, USA). Its main components were an L-7100 mobile phase pump, an L-7485 fluorescence detector and an L-761 on-line degasser. The entire system was computer-controlled with D-7000 HPLC Software Manager. All separations were made using a Zorbax Eclipse PAH column with a 4.6 mm length, 250 mm diameter and 5 μm average particle diameters. All sample injections were held constant at 20 μL using a fixed-volume injection loop. Laboratory reagent blanks were run with each series of samples under identical experimental conditions. Supplemantal Information Table S1 shows the wavelengths used during the HPLC analysis.

3.4. Solution Preparation for HPLC Analytical Figures of Merit

Calibration curves were constructed with synthetic mixtures containing DBT, BbN12T, BbN21T, C45T, BbN23T and DiN1212T at various concentrations. All the mixtures were made in 100% ACN to match the mobile phase selected for HPLC. This was accomplished by evaporation of the n-alkane PASH solvent under a gentle stream of nitrogen gas and reconstitution with ACN.

3.5. Preparation for SPE-HPLC Analytical Figure of Merit

Synthetic mixtures containing DBT, BbN12T, BbN21T, C45T, BbN23T and DiN1212T were prepared for the experiments with PASHs concentrations varying from 25 to 100 parts per trillion. The solvent composition of all the mixtures was 99.5% H_2O/0.5% ACN. This was accomplished by spiking 100 mL of H_2O with microliter volumes of PASHs solutions in ACN. SPE followed the procedure outlined in Supplemantal Information Figure S3.

3.6. Sorption Experiments

Synthetic mixtures for these experiments contained BbN12T, BbN23T and BbN21T, with final concentrations at the 50 pg/mL level. Their solvent composition was 99.5% H_2O/0.5% ACN. This was accomplished by spiking 100 mL of H_2O with microliter volumes of PASHs solutions in ACN. Six (6) pellets of each plastic consisting of approximately 150 mg of Nylon 11 and 100 mg of Nylon 6,12 were added to 100 mL of the aqueous solution and shaken for various exposure times, which ranged from 0.5 h to 4 h. An SPE-HPLC analysis of the aqueous solution was conducted at 0.5, 1, 1.5, 2, 2.5, 3, 3.5 and 4 h. An SPE-HPLC analysis of synthetic mixtures shaken in the absence of microplastics was run in parallel at all time intervals.

4. Conclusions

The SPE-HPLC method developed in these studies makes it possible to determine PASH in aqueous solutions at pg/mL concentration levels. To reach such low concentration levels, excitation and fluorescence wavelengths were optimized for each of the studied fluorophores. By doing so, excellent analytical figures of merit were obtained at relevant environmental concentrations. With the new method, the sorption uptake of benzonaphthothiophene isomers onto nylon microplastics was demonstrated for the first time. In all cases, the nylon type and the isomer structure played a role in the sorption uptake, which followed pseudo-second-order kinetics. Our findings, therefore, indicate that the interaction of PASH with microplastics in aquatic environments deserves more attention from both the remediation and the toxicological perspectives.

Supplementary Materials: The following supporting information can be downloaded at: https://www.mdpi.com/article/10.3390/molecules29071653/s1, Figure S1: HPLC-FL Chromatogram of the PASH. All PASH concentration is 50 ng/mL. Table S1: Summary of Fluorescence wavelengths used for HPLC Detection; Figure S2: Summary of SPE method used in this study; Figure S3: HPLC-FL Chromatogram following SPE extraction procedure, where the concentration of all PASH were 25 pg/mL prior to extraction.; Table S2: Values used for the calculation of the maximum sorption uptake with corresponding standard deviations; Equation (S1)–(S3) Error Propagation; Table S3: Maximum Sorption Uptake Masses and Statistical Comparison, Figure S4: Structures of PASH used in this study; Figure S5: Structures of Nylon 11 and Nylon 6,12 used in this study; Table S4: Physicochemical Properties for Nylon 11 and Nylon 6,12; Table S5: Dimensions of 10 Randomly Selected Nylon Pellets.

Author Contributions: S.D.N.: conduction of research via performance of experiments; data collection; data analysis; first draft writing. A.D.C.: formulation of research goals and aims; development of methodology; management of research activity planning and execution; writing review. All authors have read and agreed to the published version of the manuscript.

Funding: This research received no external funding.

Institutional Review Board Statement: Not applicable.

Informed Consent Statement: Not applicable.

Data Availability Statement: Data is contained within this article and Supplementary Materials.

Conflicts of Interest: The authors declare no conflict of interest.

References

1. Arthur, C.; Baker, J.; Bamford, H. (Eds.) Proceedings of the International Research Workshop on the Occurrence, Effects and Fate of Micro-Plastic Marine Debris, 9–11 September 2008. NOAA Technical Memorandum NOS-OR&R-30. 2009. Available online: https://marine-debris-site-s3fs.s3.us-west-1.amazonaws.com/s3fs-public/publications-files/TM_NOS-ORR_30.pdf?VersionId=AkHQs2er_rm6MTlJLwSTu35mauQxDuaU (accessed on 12 October 2019).
2. Wagner, J.; Wang, Z.-M.; Ghosal, S.; Rochman, C.; Gassel, M.; Wall, S. Novel method for the extraction and identification of microplastics in ocean trawl and fish gut matrices. *Anal. Methods* **2017**, *9*, 1479–1490. [CrossRef]
3. Gao, F.; Lia, J.; Sun, C.; Zhang, L.; Jiang, F.; Cao, W.; Zheng, L. Study on the capability and characteristics of heavy metals enriched on microplastics in marine environment. *Mar. Pollut. Bull.* **2019**, *144*, 61–67. [CrossRef] [PubMed]
4. Costigan, E.; Collins, A.; Hatinoglu, M.D.; Bhagat, K.; MacRae, J.; Perreault, F.; Apul, O. Adsorption of organic pollutants by microplastics: Overview of a dissonant literature. *J. Hazard. Mater. Adv.* **2022**, *6*, 100091. [CrossRef]
5. Wang, Y.; Yang, Y.; Liu, X.; Zhao, J.; Liu, R.; Xing, B. Interaction of Microplastics with Antibiotics in Aquatic Environment: Distribution, Adsorption, and Toxicity. *Environ. Sci. Technol.* **2021**, *55*, 15579–15595. [CrossRef] [PubMed]
6. Gerdes, Z.; Ogonowski, M.; Nybom, I.; Ek, C.; Adolfsson-Erici, M.; Barth, A.; Gorokhova, E. Microplastic-mediated transport of PCBs? A depuration study with Daphnia magna. *PLoS ONE* **2019**, *14*, e0205378. [CrossRef] [PubMed]
7. Qui, Y.; Zheng, M.; Wang, L.; Zhao, L.; Lou, Y.; Shi, L.; Qu, L. Sorption of polyhalogenated carbazoles (PHCs) to microplastics. *Mar. Pollut. Bull.* **2019**, *146*, 718–728. [CrossRef]
8. Wang, F.; Zhang, M.; Sha, W.; Wang, Y.; Hao, H.; Dou, Y.; Li, Y. Sorption Behavior and Mechanisms of Organic Contaminants to Nano and Microplastics. *Molecules* **2020**, *25*, 1827. [CrossRef]
9. NOAA, Technical Memorandum NMFS-NWFSC-125 Northwest Fisheries Science Center's Analyses of Tissue, Sediment, and Water Samples for Organic Contaminants by Gas Chromatography/mass Spectrometry and Analyses of Tissue for Lipid Classes by Thin Layer Chromatography/flame Ionization Detection. Washington, DC, USA. 2014. Available online: https://repository.library.noaa.gov/view/noaa/4626 (accessed on 12 October 2019).
10. European Commission, Priority Substances and Certain Other Pollutants According to Annex II of Directive 2008/105/EC. 2019. Available online: https://ec.europa.eu/environment/water/water-framework/priority_substances.htm (accessed on 12 October 2021).
11. EPA United States Environmental Protection Agency. Priority Pollutant List. 2014. Available online: https://www.epa.gov/sites/default/files/2015-09/documents/priority-pollutant-list-epa.pdf (accessed on 12 October 2021).
12. Eastmond, D.A.; Booth, G.M.; Lee, M.L. Toxicity, Accumulation, and Elimination of Polycyclic Aromatic Sulfur Heterocycles in Daphnia magna. *Arch. Environ. Contam. Toxicol.* **1984**, *13*, 105–111. [CrossRef]

13. Siemers, A.K.; Mänz, J.; Wolf-Ulric, P.; Ruck, W. Development and application of a simultaneous SPE-method for polycyclic aromatic hydrocarbons (PAHs), alkylated PAHs, heterocyclic PAHs (NSO-HET) and phenols in aqueous samples from German Rivers. *Chemosphere* **2015**, *122*, 105–114. [CrossRef] [PubMed]
14. MacKenzie, M.J.; Hunter, J.V. Sources and fates of aromatic compounds in urban stormwater runoff. *Environ. Sci. Technol.* **1979**, *13*, 179–183. [CrossRef]
15. Andrady, A.L. Microplastics in the marine environment. *Mar. Pollut. Bull.* **2011**, *62*, 1596–1605. [CrossRef] [PubMed]
16. Ambeed. Available online: https://www.ambeed.com/products/205-43-6.html (accessed on 11 January 2024).
17. Chemical Properties of Benzo[b]naphtho[2,3-d]thiophene. Chemeo. Available online: https://www.chemeo.com/cid/66-382-5/Benzo-b-naphtho-2-3-d-thiophene (accessed on 11 January 2024).
18. Chemical Properties of Benzo[b]naphtho[2,1-d]thiophene. Chemeo. Available online: https://www.chemeo.com/cid/42-561-2/Benzo-b-naphtho-2-1-d-thiophene (accessed on 11 January 2024).
19. Farkas, O.; Heberger, K.; Zenkevich, I.G. Quantitative structure-retention relationships. XIV. Prediction of gas chromatographic retention indices for saturated O-, N-, and S heterocyclic compounds. *Chemometr. Intell. Lab. Syst.* **2004**, *72*, 173–184. [CrossRef]
20. Wilson, W.B.; Hayes, H.V.; Sander, L.C.; Campiglia, A.D.; Wise, S.A. Normal-phase liquid chromatography retention behavior of polycyclic aromatic sulfur heterocycles and alkyl-substituted polycyclic aromatic sulfur heterocycle isomers on an aminopropyl stationary phase. *Anal. Bioanal. Chem.* **2018**, *410*, 1511–1524. [CrossRef] [PubMed]
21. Sahoo, T.R.; Prelot, B. Chapter 7—Absorption Processes for the removal of contaminants from wastewater: The perspective of nanomaterials and nanotechnology. In *Nanomaterials for the Detection and Removal of Wastewater Pollutants*, 1st ed.; Bonelli, B., Freyria, F.S., Rossetti, I., Sethi, R., Eds.; Elsevier: Amsterdam, The Netherlands, 2020; pp. 161–222. ISBN 9780128184905.
22. Nylon 11. Available online: https://www.sigmaaldrich.com/US/en/product/aldrich/181153#product-documentation (accessed on 17 March 2024).
23. Nylon 6,12. Available online: https://www.sigmaaldrich.com/US/en/product/aldrich/181145 (accessed on 17 March 2024).
24. Baldi, L.D.C.; Iamazaki, E.T.; Atvars, T.D.Z. Evaluation of the polarity of polyamide surfaces using the fluorescence emission of pyrene. *Dye. Pigment.* **2008**, *76*, 669–676. [CrossRef]

molecules

MDPI

Article

Multiclass Analysis for the Determination of Pharmaceuticals and Their Main Metabolites in Leafy and Root Vegetables

Carmen Mejías, Marina Arenas, Julia Martín *, Juan Luis Santos, Irene Aparicio and Esteban Alonso

Departamento de Química Analítica, Escuela Politécnica Superior, Universidad de Sevilla, C/Virgen de África 7, E-41011 Seville, Spain; cmpadilla@us.es (C.M.); mamolina@us.es (M.A.); jlsantos@us.es (J.L.S.); iaparicio@us.es (I.A.); ealonso@us.es (E.A.)

* Correspondence: jbueno@us.es

Abstract: The irrigation of soils with reclaimed contaminated wastewater or its amendment with sewage sludge contributes to the uptake of pharmaceuticals by vegetables growing in the soil. A multiresidue method has been devised to determine five pharmaceuticals and nine of their main metabolites in leafy and root vegetables. The method employs ultrasound-assisted extraction, clean-up via dispersive solid-phase extraction, and analysis through liquid chromatography–tandem mass spectrometry. Box–Behnken design was used to refine variables such as extraction solvent volume, time of extraction, number of extraction cycles, and the type and amount of d-SPE sorbent. The method achieved linearity (R^2) greater than 0.994, precision (relative standard deviation) under 16% for most compounds, and detection limits ranging from 0.007 to 2.25 ng g^{-1} dry weight. This method was applied to a leafy vegetable (lettuce) and to a root vegetable (carrot) sourced from a local market. Parent compounds were detected at higher concentrations than their metabolites, with the exception of carbamazepine-10,11-epoxide.

Keywords: pharmaceuticals; metabolites; vegetables; LC-MS/MS; carrot; lettuce

1. Introduction

Vegetables that grow in agricultural soils irrigated with contaminated reclaimed wastewater or amended with contaminated sludge or compost from wastewater treatment plants can uptake contaminants through their roots [1–3]. Reclaimed wastewater irrigation is widely used nowadays and applying sludge to soils offers agronomic advantages due to its content in organic matter [4]. However, unfortunately, wastewater treatments plants are not designed to eliminate emerging contaminants [5]. Consequently, these pollutants can accumulate in crop soils [6]. Once in the soil, pollutants can be taken up by plants, accumulate in roots, or translocate within the plant [7–11]. Accumulating contaminants in leaves and roots can negatively affect crops and could pose potential human health risks, especially when affecting edible plants [12,13]. The assessment of plant uptake requires accurate and sensitive analytical methodologies [14]. In recent years, these methods focused on detecting pesticides in vegetables [15]. Pharmaceuticals' active compounds are one of the most studied emerging pollutants. However, analytical methods for detecting pharmaceuticals metabolites in vegetables are limited and mainly focus on the parent compound, with less attention to their metabolites. Some of their metabolites can be more toxic, persistent, and concentrated in the aquatic environment than the parent compounds [16]. Various extraction methodologies have been reported for detecting pharmaceuticals in vegetables, including ultrasound-assisted extraction (UAE) [17,18], pressurised solvent extraction (PLE) [19,20], matrix solid-phase dispersion (MSPD) [21], and QuEChERS (quick, easy, cheap, effective, rugged, and safe) method [22,23]. Extract clean-up is mainly performed using solid-phase extraction (SPE) [17,18,20,24] or dispersive solid-phase extraction (d-SPE) [25]. However, sample treatment generally involves solid–

liquid extraction and SPE clean-up, followed by liquid chromatography–tandem mass spectrometry (LC–MS/MS) [25] or gas chromatography–mass spectrometry (GC–MS) [26].

The objective of this study was to develop and validate a multiclass analytical methodology for determining a broad range of pharmaceuticals (five parent compounds from four therapeutic groups) and nine of their main metabolites in leafy and root vegetables. The selected parent compounds, which included caffeine (CAF), diclofenac (DIC), and ibuprofen (IBU), were chosen based on their prevalent usage and consumption; their inefficient removal during wastewater treatment processes, leading to their widespread presence in various environmental compartments; and their potential ecotoxicological risks, as observed with compounds like carbamazepine (CBZ) and sulfamethoxazole (SMX). To the best of our knowledge, this is the first method to quantify pharmaceutical metabolites for different therapeutic groups in vegetables.

2. Results and Discussion

2.1. Method Optimisation

The method was optimised with spiked carrot samples (0.5 g) at 100 ng g^{-1} dry weight (dw). Samples were individually spiked with the compounds using a methanol solution to ensure thorough permeation of the entire sample. The samples were then vortex-mixed for homogenisation and incubated in the dark for 12 h to allow for proper evaporation of the solvent and equilibration. For the optimisation of d-SPE sorbent, the spiked procedure took place after extraction and before the addition of the clean-up sorbent.

2.1.1. Optimisation of the Extraction Solvent

Extraction solvents tested were acetonitrile (ACN), acetone, and hexane. Methanol (MeOH) was initially included as a tested solvent; however, methanol extracts were characterised by the highest extraction of matrix components, making the handling of the extracts difficult and, thus, they were not analysed further. Samples were extracted three times in an ultrasonic bath for 10 min, centrifuged at 2900× g for 10 min, and subjected to clean-up by d-SPE with 0.8 g of C18. Optimisation was carried out in triplicate with 3 mL of the tested solvent. Extraction recovery was calculated by comparison of the signal of the spiked vegetable with a matrix-matched standard at the same spiking concentration. As can be seen in Figure 1a, the best extraction recoveries were obtained with acetone, so acetone was preselected.

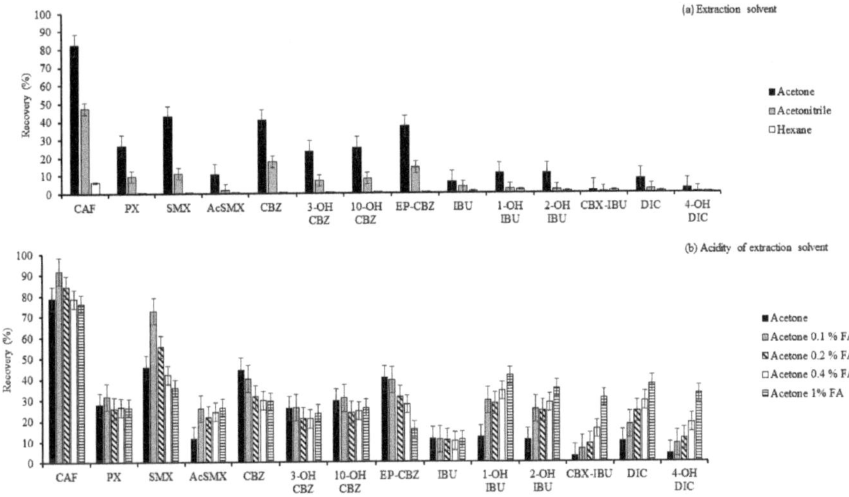

Figure 1. Optimisation of (**a**) extraction solvent and (**b**) acidity of extraction solvent (n = 3).

Then, different percentages of acidity of acetone were tested (acetone, acetone (0.1% v/v, formic acid), acetone (0.2% v/v, formic acid), acetone (0.4% v/v, formic acid), and acetone (1% v/v, formic acid)). The extraction procedure was the same as in the optimisation of the solvent. Extraction recoveries were also obtained in the same way as in the previous experiment. As can be seen in Figure 1b, the best extraction recoveries for most compounds, especially for those with lower recoveries, were obtained with acetone (1% v/v, formic acid), so this solvent was selected as extraction solvent. IBU and metabolites were better extracted at pH < pKa, as they were not ionised at these pH values. The possible evaporation of acetone during the extraction process was evaluated, obtaining no differences in volume.

2.1.2. Optimisation of d-SPE Sorbents and Their Amount

Clean-up was optimised to select the most appropriate sorbent or sorbents for removal of interfering compounds without removing target compounds. Three clean-up sorbents were tested: a weak anion exchanger sorbent (primary-secondary amine, PSA), a reverse phase sorbent (C18), and a normal phase sorbent (florisil). PSA sorbent is used to remove polar pigments, fatty acids, sugar, and organic acids; C18 is commonly used to remove nonpolar and moderately polar compounds, such as lipophilic compounds; florisil is suitable to remove polar compounds. A Box–Behnken design (BBD) was applied to optimise d-SPE sorbent amount. The variables (types of sorbents: C18, PSA and florisil) were evaluated at three levels (mass of sorbent: 0, 0.4, and 0.8 g). The number of experiments (N) required is defined by the equation: $N = 2k (k - 1) + C_0$, where k is the number of variables and C_0 is the number of central points. In this case, k and C_0 values were set at 3, resulting in a total of 15 experiments. Samples were extracted three times in an ultrasonic bath for 10 min with 3 mL of acetone (1% v/v, formic acid) and centrifuged at 2900× g for 10 min. Liquid phases of vegetable matrix extractions were spiked at 200 ng mL^{-1} and subjected to the 15 different clean-up experiments. The experiments were randomly performed to minimize the effects of uncontrolled variables. The BBD matrix is shown in Supplementary Materials Table S1. BBD values were calculated using Statgraphics 18-X64 for Windows, using the optimisation of multiple responses mode for the analysis of the results. Clean-up efficacy was evaluated to select the most appropriated sorbent. Higher efficacy is achieved when the tested sorbent removes interferences to a greater extent (inducing less matrix effect) while it does not remove target compounds. Clean-up efficacy was calculated by comparison of the signal of a matrix-matched standard with a standard in pure solvent at the same spiking concentration. Figure 2 shows response surface plots corresponding to clean-up efficacy (%) vs. C18 amount (g) and PSA amount (g) (Figure 2a), clean-up efficacy (%) vs. C18 amount (g) and florisil amount (g) (Figure 2b), and clean-up efficacy (%) vs. PSA amount (g) and florisil amount (g) (Figure 2c). The best values of clean-up efficacy (%) were obtained for 0 g of PSA, 0.8 g of C18, and 0.15 g of florisil. According to the results, 0.8 g of C18 and 0.15 g of florisil were selected as d-SPE sorbents amount for clean-up. Due to its chemical structure, the C18 sorbent was able to remove nonpolar interferences from the extract without removing the selected compounds, making it a suitable sorbent for clean-up.

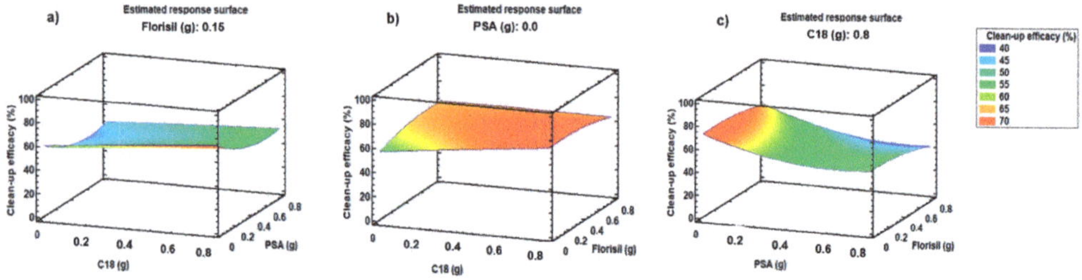

Figure 2. Response surface plots, corresponding to clean-up efficacy (%) of target compounds, versus (**a**) C18 amount (g) and PSA amount (g); (**b**) C18 amount (g) and florisil® amount (g) (**c**); PSA amount (g) and florisil® amount (g).

2.1.3. Optimisation of Extraction Solvent Volume, UAE Time, and Number of Extraction Cycles

As in the optimisation of d-SPE, a BBD was applied to optimise extraction solvent volume, UAE time, and number of extraction cycles. Three levels were evaluated for each variable: acetone (1% v/v, formic acid) volume (3, 4.5, and 6 mL), UAE time (5, 10, and 15 min), and number of extraction cycles (1, 2, and 3). As in the previous case, 15 experiments were carried out. Details of each experiment can be seen in Table S2 in Supplementary Materials. Samples were subjected to 15 extraction experiments and were cleaned-up with 0.8 g of C18 and 0.15 g of florisil according to the optimisation of d-SPE sorbent and their amount. Response surface plots corresponding to method of recovery were constructed to better evaluate the effects of the extraction method on each variable and their interactions. BBD values were calculated using Statgraphics 18-X64 for Windows, using the optimisation of multiple responses mode for the analysis of the results. Figure 3 shows the response surface plots corresponding to the method of recovery (%) vs. acetone (0.1% v/v, formic acid) volume and UAE time (Figure 3a), method of recovery (%) vs. acetone (0.1% v/v, formic acid) volume and number of extraction cycles (Figure 3b), and method of recovery (%) vs. number of extraction cycles and UAE time (Figure 3c). Response surface plots facilitate the optimum value within the evaluated range of each variable, but do not have to coincide exactly with the three values tested in the experiments. As can be seen in Figure 3, number of extraction cycles was the most influential parameter. The best values of method of recovery (%) were obtained for 4 mL of acetone (1% v/v, formic acid), three extraction cycles, and 9 min of UAE. According to the results, 4 mL of acetone (1% v/v, formic acid), three extraction cycles, and 9 min of extraction were selected for sample extraction. As can be seen from the results obtained, the number of extraction cycles proved to be the most influential factor in the extraction of the analytes. The extraction time and the volume of solvent ended up having very little influence on the extraction, with it being more of a priority to repeat the process several times than to add a large volume of solvent and to prolong the extraction for a long time.

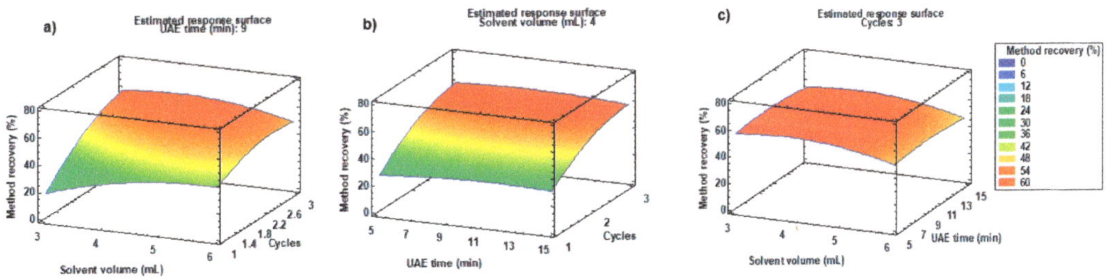

Figure 3. Response surface plots, corresponding to method of recovery (%) of target compounds, versus (**a**) acetone (0.1% v/v, formic acid) volume (mL) and number of extraction cycles; (**b**) UAE time (min) and number of extraction cycles; and (**c**) acetone (0.1% v/v, formic acid) volume (mL) and UAE time (min).

2.2. Method Validation

The method was validated in terms of sensitivity (method detection limits (MDL) and method quantification limits (MQL)), linearity, accuracy (A), and precision. Previously, matrix effect (ME) was evaluated by comparison of calibration curve slopes in pure solvent (external calibration curves) and calibration curve slopes in sample extract (matrix-matched calibration curves). Calibration curves were prepared in triplicate at five concentration levels in the range from 2.25 to 100 ng g^{-1} dw and were constructed using analyte area (axis y) versus analyte concentration (axis x). Student's *t*-test, at 95% of confidence, revealed statistical differences between slopes so matrix-matched calibration curves were applied for quantification.

The methodology was validated for leafy and root vegetables by using spiked lettuces and carrots. In order to evaluate the linearity, five-point matrix-matched calibration curves were prepared in triplicate in the range from MQL to 100 ng g^{-1}. Correlation coefficients (R^2) were equal or higher than 0.994 for all the compounds in both matrices (Table 1). Instrument detection limit (IDL) and quantification limit (IQL) values were estimated as the concentrations corresponding to signal-to-noise ratios of 3 and 10, respectively, by injecting spiked extracts at low concentration levels. MDL and MQL were calculated from IDL and IQL, applying the concentration factor, matrix effect, and recovery for each compound. MDL values were in the range from 0.002 ng g^{-1} to 1.020 ng g^{-1} dw in lettuces and from 0.007 ng g^{-1} to 0.417 ng g^{-1} dw in carrots. MQL values were from 0.007 ng g^{-1} to 2.25 ng g^{-1} dw in lettuces and from 0.04 ng g^{-1} to 2.25 ng g^{-1} dw in carrots (Table 1).

Table 1. Method detection limits (MDL), method quantification limits (MQL), and matrix-matched calibration curve correlation coefficients (R^2).

Compound	Lettuce			Carrot		
	MDL (ng g^{-1} dw)	MQL (ng g^{-1} dw)	R^2	MDL (ng g^{-1} dw)	MQL (ng g^{-1} dw)	R^2
CAF	0.011	0.027	0.998	0.022	0.217	0.995
PX	0.013	0.313	0.998	0.018	0.045	0.997
CBZ	0.003	0.008	0.998	0.017	0.050	0.995
3-OH CBZ	0.010	0.052	0.998	0.027	0.068	0.995
10-OH CBZ	0.004	0.012	0.997	0.018	0.053	0.995
EP-CBZ	0.003	0.010	0.995	0.020	0.059	0.995
DIC	0.009	0.086	0.999	0.016	0.040	0.996
4-OH DIC	0.002	0.007	0.997	0.039	0.781	0.994
IBU	0.833	2.25	0.997	0.833	2.25	0.996
1-OH IBU	0.164	0.820	0.998	0.179	0.446	0.996
2-OH IBU	0.003	0.009	0.998	0.023	2.326	0.998
CBX-IBU	1.020	2.25	0.999	0.018	1.761	0.998
SMX	0.008	0.024	0.997	0.417	1.667	0.996
AcSMX	0.007	0.070	0.999	0.049	0.196	0.996

Parent compounds are marked in bold.

Recovery and precision were evaluated from spiked samples at three concentration levels (2.5, 40, and 100 ng g^{-1} dw) in triplicate. Recovery (%) was calculated as the percentage of analyte extracted. This percentage was quantified using the signal of a matrix-matched standard as reference to avoid the influence of matrix effect as follows: $R\ (\%) = (A_{\text{spiked sample}} - A_{\text{non-spiked sample}}) \times 100/(A_{\text{spiked extract}} - A_{\text{non-spiked sample}})$. Recovery for most compounds was in the range from 28% to 98% for lettuces and in the range from 26% to 115% for carrots (Table 2). For some analytes, recovery values were higher at the low concentration level; however, the low levels of concentration that were tested (ppt in all cases) could explain this variation. Concerning recovery values, analytical guidelines, like those from the AOAC Peer-Verified Methods program [27], suggest that recoveries can fall between 60% and 120% at part per billion levels, typically referring to "absolute recovery" or "accuracy". Given that the method proposed in this work is a multi-analyte approach, where compounds with highly varied properties are analysed, it is necessary to achieve a balance that ensures optimal recovery for all analytes. Furthermore, as shown in Table 3, the accuracy reached values between 81% and 107% for all analytes. This parameter reflects the suitability of the method. Precision, expressed as relative standard deviation (% RSD), was determined from the analysis of spiked samples in triplicate on two different days. RSD values were below 16% for most compounds in both types of matrices at the three spiked concentration levels for all compounds. As expected, precision was lower at lower spiking concentrations, because the lower the concentration, the greater the dispersion obtained in the extraction data.

Table 2. Recovery (%) and precision, expressed as relative standard deviation (RSD%), for lettuce and carrot matrices at three spiking levels.

Compound	Lettuce						Carrot					
	2.5 (ng g^{-1} dw)		40 (ng g^{-1} dw)		100 (ng g^{-1} dw)		2.5 (ng g^{-1} dw)		40 (ng g^{-1} dw)		100 (ng g^{-1} dw)	
	R (%)	RSD (%)	R (%)	RSD (%)	R (%)	RSD (%)	R (%)	RSD (%)	R (%)	RSD (%)	R (%)	RSD (%)
CAF	94	12	34	0	37	11	115	10	41	4	33	7
PX	80	21	49	5	28	6	55	15	41	6	38	9
CBZ	66	22	47	9	55	7	20	24	19	1	27	8
3-OH CBZ	48	16	48	7	51	8	37	3	52	10	47	14
10-OH CBZ	43	18	45	8	51	10	19	25	17	4	26	8
EP-CBZ	48	18	20	1	21	10	17	18	26	24	19	3
DIC	58	25	24	16	27	6	62	23	24	14	21	3
4-OH DIC	67	23	61	2	38	4	64	12	34	18	31	7
IBU	13	27	23	1	31	7	20	30	48	21	60	28
1-OH IBU	61	28	45	9	53	9	56	7	48	9	40	11
2-OH IBU	55	14	44	5	41	7	43	18	42	11	37	9
CBX-IBU	98	16	43	5	52	9	142	19	48	16	42	13
SMX	21	30	21	3	25	8	16	22	26	17	27	2
AcSMX	71	30	51	6	45	9	51	11	44	14	42	13

Parent compounds are marked in bold.

Table 3. Accuracy (%) and matrix effect (%) for lettuce and carrot matrices at three spiking levels.

Compound	Lettuce						Carrot					
	2.5 (ng g^{-1} dw)		40 (ng g^{-1} dw)		100 (ng g^{-1} dw)		2.5 (ng g^{-1} dw)		40 (ng g^{-1} dw)		100 (ng g^{-1} dw)	
	A (%)	ME (%)	A (%)	ME (%)	A (%)	ME (%)	A (%)	ME (%)	A (%)	ME (%)	A (%)	ME (%)
CAF	87	−27.6	95	−34.8	102	−0.34	81	−62.4	99	−37.6	107	−35.0
PX	88	−13.2	94	−19.9	103	−0.22	84	−54.2	95	−64.3	99	−27.1
CBZ	86	−7.07	92	−1.77	101	−0.01	82	−28.8	96	−0.81	98	−6.33
3-OH CBZ	89	−15.9	93	−11.9	100	−0.10	83	−44.8	98	−24.4	100	−10.6
10-OH CBZ	81	−7.41	96	−5.00	99	−0.03	85	−2.73	97	−7.37	102	−2.93
EP-CBZ	84	−30.8	91	−12.1	97	−0.10	87	−11.2	100	−19.8	104	−8.05
DIC	82	−15.3	91	−39.0	98	−0.31	88	−53.5	92	−55.2	107	−15.8
4-OH DIC	83	−32.0	94	−11.7	100	−0.10	86	−6.27	94	−26.02	105	−4.11
IBU	88	−11.2	93	−8.76	101	−0.32	89	−35.7	93	−22.2	103	−8.41
1-OH IBU	87	−9.76	95	−4.92	104	−0.02	91	−28.7	91	−0.08	106	−0.42
2-OH IBU	87	−6.06	98	−4.31	106	−0.04	92	−45.6	100	−27.7	101	−6.85
CBX-IBU	85	−7.98	96	−0.45	101	−0.01	90	−49.5	97	−6.49	97	−1.24
SMX	84	−20.3	94	−20.8	99	−0.19	85	−79.3	98	−75.7	99	−26.1
AcSMX	81	−1.81	97	0.02	98	−0.01	82	−33.4	99	−29.9	101	−1.32

Parent compounds are marked in bold.

Matrix effect was calculated as follows: ME (%) = ($A_{\text{spiked extract}} - A_{\text{non-spiked extract}} - A_{\text{standard}}$) × 100/$A_{\text{standard}}$. Matrix effect ranged from −0.01% to −39.0% in leafy vegetables and from −0.08% to −79.3% in root vegetables (Table 3). In all cases, matrix ion suppression occurred.

Accuracy was obtained using the following equation: A (%) = ($C_{\text{spiked sample}} - C_{\text{non-spiked sample}}$) × 100/$C_{\text{spike concentration}}$. Accuracy percentage ranged from 81% to 107% (Table 3).

In Figure S1 in Supplementary Materials, a chromatogram of a spiked lettuce (10 ng g^{-1} dw) can be seen.

2.3. Method Comparison

As can be seen from Table 4, the therapeutic group most studied in vegetables was antibiotics, as well as in the assessment of plant uptake. Most of the methods used 0.5 g of sample amount, but some of QuEChERS methods [22,23] used amounts up to 10 g. Regarding time required for extraction, most methodologies are comparable in duration to the method developed in this work, since several extraction cycles are often necessary for good performance and techniques such as QuEChERS include different procedures. There are just two methodologies that require a significantly shorter time than the one proposed [17,20]. Comparing the type and amount of solvent used, all methods use higher amounts and/or more toxic solvents than acetone, such as ACN, hexane, or MeOH [28]. There are only three methods that require less than 10 mL of extraction solvent [20,22,23], but at the same time using mixtures of ACN and MeOH, and just one not employing organic solvents [17]. All methodologies carry out some clean-up procedure to remove interfering compounds that may affect the analytical determination. Some of them incorporate this clean-up in the same step as the extraction [19,20,22,23], usually in QuEChERS techniques. In other cases, including the proposed methodology, it is performed as an additional stage [17,18,20,24], which, although it adds an extra step to the method, helps to eliminate interferences and to increase the sensitivity and selectivity of the method. Therefore, in terms of green sample preparation, the proposed method can be considered one of the most sustainable when compared to similar reported methodologies for analysing pharmaceutical residues in vegetables. Only one methodology [19] used GC-MS for the determination of the compounds, while the other ones employed LC-MS/MS. The proposed analytical method reported the lowest MQLs (highest sensitivity) in comparison to reported methods in the literature.

Table 4. Comparison with other methods reported in the literature.

Compounds	Sample Amount (g)	Extraction Technique	Time (min)	Solvent (mL)	Clean-Up Step	Analytical Determination	MQL (ng g^{-1} dw)	Reference
1 CAF, 1 analgesic, 8 antibiotics, and 1 anticonvulsant	0.5	QuEChERS	3.5	ACN, MeOH 5	-	LC-MS/MS	0.7–8.0 (MDL)	[20]
1 CAF, 1 analgesic, 8 antibiotics, and 1 anticonvulsant	0.5	PLE	20	ACN, MeOH 40	SPE	LC-MS/MS	1.9–15.8 (MDL)	[20]
7 antibiotics	0.25–0.5	UAE	6	18 (Alkaline solution with Mg^{2+})	SPE	LC-MS/MS	2–10	[17]
1 anticonvulsant	0.5	PLE	27	n.d. (Acetone, hexane)	-	GC-MS	7.6–61.7	[19]
11 antibiotics	1.0	UAE	30	20 (ACN)	SPE	LC-MS/MS	0.2–6.25	[18]
10 antibiotics and 6 of their metabolites	1.0	UAE	30	20 (MeOH)	SPE	LC-MS/MS	0.2–9.2	[24]
26 antibiotics	10	QuEChERS	27	10 (ACN, MeOH)	-	LC-MS/MS	0.02–1.5	[23]
20 antibiotics	10	QuEChERS	35	10 (ACN, MeOH)	-	LC-MS/MS	0.33–2.92 (MDL)	[22]
1 CAF, 2 NSAIDs, 1 antibiotic, 1 anticonvulsant, and 9 of their main metabolites	0.5	UAE	27	12 (Acetone)	d-SPE	LC-MS/MS	0.007–2.25	Proposed methodology

n.d.: no data.

2.4. Method Application

The method was applied to the determine target compounds in one type of leafy vegetable (lettuce) and one type of root vegetable (carrot). Three carrot samples and three lettuce samples were purchased from local markets and analysed. Concentrations are shown in Supplementary Materials (Table S3). SMX was the compound with the highest detection frequency and highest concentration (up to 7.8 ng g^{-1} dw in carrots). In contrast, 2-hydroxyibuprofen (2-OH-IBU), carboxyibuprofen (CBX-IBU), and 4-hydroxydiclofenac (4-OH DIC) were not detected in any of the analysed samples. In general, parent compounds were found at higher concentrations than their metabolites, with exception of carbamazepine-10,11-epoxide (EP-CBZ). These results show the applicability of the method. The concentrations levels measured were similar to or lower than those reported by other authors for pharmaceuticals in vegetables. As an example, Calderón-Preciado et al. (2009) [19] reported IBU concentration levels in lettuce vegetables of 28.5 ng g^{-1}. To our knowledge, this is the first time that concentration levels have been reported for metabolites in lettuce and carrots.

3. Materials and Methods

3.1. Chemicals and Reagents

Analytical standards of 1,3,7-trimethylxanthine (CAF), 1,7-dimethylxanthine (paraxanthine) (PX), CBZ, 3-hydroxycarbamazepine (3-OH CBZ), 10,11-dihydro-10-hydroxycarba mazepine (10-OH CBZ), EP-CBZ, DIC, 4-OH DIC, IBU, 1-hydroxyibuprofen (1-OH IBU), 2-OH IBU, CBX-IBU, SMX, and N^4-acetylsulfamethoxazole (AcSMX) were supplied by Sigma-Aldrich (Steinheim, Germany). Physical–chemical properties of the target compounds can be seen in Table S4 in Supplementary Materials. Caffeine-$^{13}C_3$ (CAF-$^{13}C_3$) and ibuprofen-d_3 (IBU-d_3) used as internal standards were purchased from Sigma-Aldrich (Steinheim, Germany). Individual stock standard solutions were prepared at 1000 μg mL^{-1} in MeOH and stored in amber glass bottles at −18 °C. Working solutions were prepared by dilution of the individual standard solutions or by mixing the individual standard solutions to have a mixture of the target compounds and a mixture of the internal standards (I.S.) Florisil used as d-SPE sorbent was provided by Sigma-Aldrich (Madrid, Spain). PSA and C18 were provided by Scharlab (Barcelona, Spain). HPLC-grade ACN, hexane, and acetone were supplied by Romil (Barcelona, Spain). LC-MS-grade water and MeOH were purchased from Honeywell (Seelze, Germany). Analytical-grade formic acid (98%) was provided by Panreac (Barcelona, Spain).

3.2. Sample Collection and Treatment

Root and leafy vegetables were bought in a local market. Carrots were selected as root vegetables and lettuces were selected as leafy vegetables. Fresh and clean vegetables were cut into small pieces, triturated in a blender, freeze-dried in a Cryodos-50 lyophiliser (Telstar, Terrassa, Spain), sieved (particle size < 100 μm), saved in glass bottles, and maintained at −18 °C until extraction. Treatment was based on UAE as the extraction step and d-SPE as the clean-up step. Pre-treated vegetable samples (0.5 g dw) were weighed in glass centrifuge tubes and spiked with I.S. (CAF-$^{13}C_3$ and IBU-d_3) at 125 ng g^{-1} dw. This I.S. concentration was selected to be high enough to be observed correctly in all samples to avoid as much as possible the variation of its intensity between samples, since the higher the concentration, the better signal-to-noise ratio and the precision. Samples were extracted with 4 mL of acetone (1% v/v, formic acid) by sonication for 9 min in an ultrasonic bath at 25 °C. Afterwards, the tubes were centrifuged at 2900× g for 10 min, the liquid phase was transferred to a clean tube, and the extraction procedure was repeated twice again. The liquid phase of the three extractions was combined into a clean centrifuge tube and 0.15 g of florisil and 0.8 g of C18 were added for d-SPE extract clean-up. The tubes were vigorously shaken for 1 min and centrifuged for 15 min at 2900× g. The liquid phase was transferred to another clean tube and was evaporated to dryness under a gentle nitrogen stream in the XcelVap® system (Horizon Technology, Reading, UK). The dried extract was reconstituted in 250 μL of water–MeOH solution (1:1, v/v), filtered through a 0.22 μm cellulose syringe filter, and transferred into an automatic injector vial for injection into the LC-MS/MS system.

3.3. Liquid Chromatography–Tandem Mass Spectrometry

Chromatographic determination was carried out on an Agilent 1290 Infinity II liquid chromatographic system (Agilent, Santa Clara, CA, USA) coupled to a 6495-triple quadrupole (QqQ) mass spectrometer (MS) equipped with an electrospray ionisation source (ESI). Chromatographic separation was optimised by Malvar et al. (2019) [16] and was performed on a Kinetex® Polar C18 column (50 mm × 3.0 mm i.d., 2.6 μm particle size) (Phenomenex, Torrance, CA, USA) thermostated at 35 °C and protected by a

SecurityGuard™ ULTRA C18 guard column (2 mm × 3.0 mm i.d., 2.6 μm particle size) (Phenomenex, Torrance, CA, USA). The injection volume was 10 μL. The mobile phase consisted of water containing formic acid (0.1%, *v*/*v*) (solvent A) and MeOH containing formic acid (0.1%, *v*/*v*) (solvent B) at a flow rate of 0.6 mL min^{-1}. A gradient mode was used to separate and elute analytes from the column. Gradient elution started with a linear increase of solvent B from 2% to 38% in 9.5 min (held for 2.5 min), then to 50% in 3.5 min (held for 2.5 min), and, finally, to 98% in 4 min (held for 1 min). The mobile phase returned to initial conditions by linear decrease of solvent B from 98% to 2% in 2 min and held for 5 min for re-equilibration. Total chromatographic run time was 30 min. The ESI-MS/MS conditions were as follows: capillary voltage, 4000 V; drying gas flow rate, 11 L min^{-1}; drying gas temperature, 250 °C; sheath gas flow rate, 12 L min^{-1}; sheath gas temperature, 250 °C; and nebulizer pressure; 40 psi. The mass spectrometer was operated in dynamic multiple reaction monitoring mode (dMRM). Instrument control and data acquisition were carried out with MassHunter Workstation software version 10.1 (Agilent, Santa Clara, CA, USA). Optimised LC-MS/MS parameters for each compound are given in Supplementary Materials Table S5.

4. Conclusions

A new analytical methodology has been optimised and validated for the determination of five pharmaceuticals and nine of their main metabolites in leafy (lettuce) and root (carrot) vegetables. Sample extraction and clean-up are based on UAE and d-SPE, which are low-cost, affordable, and easy-to-perform techniques. Analytical determination is carried out by LC-MS/MS analysis. MQLs values were in the range of 0.007–2.25 ng g^{-1} dw for leafy vegetables and in the range of 0.040–2.25 ng g^{-1} dw for root vegetables. Precision, expressed as relative standard deviation, was lower than 16% for most the compounds. The method was satisfactorily applied, and some pharmaceuticals were detected in both types of vegetables, whereas others were detected just in leafy or root vegetables. The proposed method can provide a useful tool to (i) assess pharmaceuticals uptake in vegetables through soil contamination, (ii) evaluate their accumulation and translocation in plants, and (iii) study human exposure, estimating daily intake of pharmaceuticals by consumption of vegetables.

Supplementary Materials: The following supporting information can be downloaded at: https://www.mdpi.com/article/10.3390/molecules29153471/s1, Figure S1: Chromatogram obtained from a spiked lettuce (10 ng g^{-1} dw); Table S1: Box–Behnken design matrix for the optimisation of clean-up sorbent amount; Table S2: Box–Behnken design matrix for the optimisation of extraction solvent volume, UAE time extraction, and number of extraction cycles; Table S3: Analytes detected in leafy and root vegetables from local markets; Table S4: Physical–chemical properties of the target compounds; Table S5: LC-MS/MS conditions and retention times for the target compounds.

Author Contributions: Conceptualisation, E.A. and I.A.; methodology, J.L.S.; validation, C.M., M.A. and J.M.; formal analysis, J.L.S.; investigation, C.M. and M.A.; resources, J.M.; data curation, J.M.; writing—original draft preparation, C.M. and M.A.; writing—review and editing, J.M. and I.A.; visualisation, J.L.S.; supervision, E.A. and I.A.; project administration, E.A.; funding acquisition, E.A. All authors have read and agreed to the published version of the manuscript.

Funding: This research was funded by MICIU/AEI/10.13039/501100011033, grant number PID2020-117641RB-I00.

Institutional Review Board Statement: Not applicable.

Informed Consent Statement: Not applicable.

Data Availability Statement: Data will be made available on request.

Acknowledgments: C. Mejías acknowledges the University of Seville for her predoctoral contract. M. Arenas acknowledges the Spanish Ministerio de Universidades for her predoctoral contract.

Conflicts of Interest: The authors declare no conflicts of interest. The funders had no role in the design of the study; in the collection, analyses, or interpretation of data; in the writing of the manuscript; or in the decision to publish the results.

References

1. Wu, X.; Dodgen, L.K.; Conkle, J.L.; Gan, J. Plant uptake of pharmaceutical and personal care products from recycled water and biosolids: A review. *Sci. Total Environ.* **2015**, *536*, 655–666. [CrossRef]
2. Nkoh-Nkoh, J.; Shang, C.; Okeke, E.S.; Ejeromedoghene, O.; Oderinde, O.; Etafo, N.O.; Mgbechidinma, C.L.; Bakare, O.C.; Meugang, E.F. Antibiotics soil-solution chemistry: A review of environmental behavior and uptake and transformation by plants. *J. Environ. Manag.* **2024**, *354*, 120312. [CrossRef]
3. Miller, E.; Nason, S.L.; Karthikeyan, K.G.; Pedersen, J.A. Root uptake of pharmaceuticals and personal care products ingredients. *Environ. Sci. Technol* **2016**, *50*, 525–541. [CrossRef]
4. Malvar, J.L.; Santos, J.L.; Martín, J.; Aparicio, I.; Alonso, E. Occurrence of the main metabolites of the most recurrent pharmaceuticals and personal care products in Mediterranean soils. *J. Environ. Manag.* **2020**, *278*, 111584. [CrossRef]
5. Santos, J.L.; Martín, J.; Mejías, C.; Aparicio, I.; Alonso, E. Pharmaceuticals and their metabolites in sewage sludge and soils: Distribution and environmental risk assessment. Emerging pollutants in sewage sludge and soils. In *Handbook of Environmental Chemistry*; Núñez-Delgado, A., Arias-Estévez, M., Eds.; Springer: Cham, Switzerland, 2023; Volume 114, pp. 19–36. [CrossRef]
6. Mejías, C.; Martín, J.; Santos, J.L.; Aparicio, I.; Alonso, E. Occurrence of pharmaceuticals and their metabolites in sewage sludge and soil: A review on their distribution and environmental risk assessment. *Trends Environ. Anal. Chem.* **2021**, *30*, e00125. [CrossRef]
7. Ahmed, M.B.M.; Rajapaksha, A.U.; Lim, J.E.; Vu, N.T.; Kim, I.S.; Kang, H.M.; Lee, S.S.; Ok, Y.S. Distribution and accumulative pattern of tetracyclines and sulfonamides in edible vegetables of cucumber, tomato and lettuce. *J. Agric. Food Chem* **2015**, *63*, 398–405. [CrossRef]
8. Azanu, D.; Mortey, C.; Darko, G.; Weisser, J.J.; Styrishave, B.; Abaidoo, R.C. Uptake of antibiotics from irrigation water by plants. *Chemosphere* **2016**, *157*, 107–114. [CrossRef] [PubMed]
9. Abril, C.; Santos, J.L.; Martín, J.; Aparicio, I.; Alonso, E. Uptake and translocation of multiresidue industrial and household contaminants in radish grown under controlled conditions. *Chemosphere* **2021**, *268*, 128823. [CrossRef]
10. Herklotz, P.A.; Gurung, P.; Heuvel, B.V.; Kinney, C.A. Uptake of human pharmaceuticals by plants grown under hydroponic conditions. *Chemosphere* **2010**, *78*, 1416–1421. [CrossRef] [PubMed]
11. Shargil, D.; Gerstl, Z.; Fine, P.; Nitsan, I.; Kurtzman, D. Impact of biosolids and wastewater effluent application to agricultural land on steroidal hormone content in lettuce plants. *Sci. Total Environ* **2015**, *505*, 357–366. [CrossRef]
12. Riemenschneider, C.; Seiwert, B.; Moeder, M.; Schwarz, D.; Reemstsma, T. Extensive transformation of the pharmaceutical carbamazepine following uptake into intact tomato plants. *Environ. Sci. Technol.* **2017**, *51*, 6100–6109. [CrossRef]
13. Malchi, T.; Maor, Y.; Tadmor, G.; Shenker, M.; Chefetz, B. Irrigation of root vegetables with treated wastewater: Evaluating uptake of pharmaceuticals and the associated human health risks. *Environ. Sci. Technol.* **2014**, *48*, 9325–9333. [CrossRef]
14. Geng, J.; Liu, X.; Li, S. Accumulation and risk assessment of antibiotics in edible plants grown in contaminated farmlands: A review. *Sci. Total Environ.* **2022**, *853*, 158616. [CrossRef] [PubMed]
15. Masiá, A.; Suárez-Varela, M.M.; Llopis-González, A.; Picó, Y. Determination of pesticides and veterinary drug residues in food by liquid chromatography-mass spectrometry: A review. *Anal. Chim. Acta* **2016**, *936*, 40–61. [CrossRef]
16. Malvar, J.L.; Santos, J.L.; Martín, J.; Aparicio, I.; Alonso, E. Routine analytical method for monitoring the main metabolites for a recurrent group of parabens and pharmaceuticals in wastewater and tap water. *Anal. Bioanal. Chem.* **2019**, *411*, 6625–6635. [CrossRef]
17. Merlo, F.; Centenaro, D.; Maraschi, F.; Profumo, A.; Speltini, A. Green and Efficient Determination of Fluoroquinolone Residues in Edible Green Fruits and Leafy Vegetables by Ultrasound-Assisted Extraction Followed by HPLC-MS/MS. *Molecules* **2022**, *27*, 6595. [CrossRef]
18. Feng, Y.; Zhang, W.J.; Liu, Y.W.; Xue, J.M.; Zhang, S.Q.; Li, Z.J. A simple, sensitive, and reliable method for the simultaneous determination of multiple antibiotics in vegetables through SPE-HPLC-MS/MS. *Molecules* **2018**, *23*, 1953. [CrossRef] [PubMed]
19. Calderón-Preciado, D.; Jiménez-Cartagena, C.; Peñuela, G.; Bayona, J.M. Development of an analytical procedure for the determination of emerging and priority organic pollutants in leafy vegetables by pressurized solvent extraction followed by GC-MS determination. *Anal. Bioanal. Chem.* **2009**, *394*, 1319–1327. [CrossRef]
20. Chuang, Y.H.; Zhang, Y.; Zhang, W.; Boyd, S.A.; Li, H. Comparison of accelerated solvent extraction and quick, easy, cheap effective, rugged and safe method for the extraction and determination of pharmaceuticals in vegetables. *J. Chromatog. A* **2015**, *1404*, 1–9. [CrossRef] [PubMed]
21. Hoff, R.B.; Futigami, L.S.; Pierezan, M.D.; Deolindo, C.T.P.; de Melo, A.P.Z.; Molognoni, L.; Pimenta, R.; Burin, V.M.; de Francisco, A.; Daguer, H. Cassava-based materials for matrix solid phase dispersion: An alternative for sample preparation in food analysis. *J. Chromatog. B* **2022**, *1201–1202*, 123263. [CrossRef]
22. Yu, X.; Liu, H.; Pu, C.; Chen, J.; Sun, Y.; Hu, L. Determination of multiple antibiotics in leafy vegetables using QuEChERS-UHPLC-MS/MS. *J. Sep. Sci.* **2017**, *41*, 713–722. [CrossRef]

23. Hu, F.; Bian, K.; Liu, Y.; Su, Y.; Zhou, T.; Song, X.; He, L. Development of a modified QUick, Easy, CHeap, Effective, Ruffed and Safe method for the determination of multi-class antimicrobials in vegetables by liquid chromatography tandem mass spectrometry. *J. Chromatog. A* **2014**, *1368*, 52–63. [CrossRef] [PubMed]
24. Tadic, D.; Matamoros, V.; Bayona, J.M. Simultaneous determination of multiclass antibiotics and their metabolites in four types of field-grown vegetables. *Anal. Bioanal. Chem.* **2019**, *411*, 5209–5222. [CrossRef]
25. Aparicio, I.; Martín, J.; Abril, C.; Santos, J.L.; Alonso, E. Determination of household and industrial chemicals, personal care products and hormones in leafy and root vegetables by liquid chromatography-tandem mass spectrometry. *J. Chromatog. A* **2018**, *1533*, 49–56. [CrossRef] [PubMed]
26. Albero, B.; Sánchez-Brunete, E.M.; Tadeo, J.L. Application of matrix solid-phase dispersion followed by GC-MS/MS to the analysis of emerging contaminants in vegetables. *Food Chem* **2017**, *217*, 660–667. [CrossRef] [PubMed]
27. AOAC International. Method Validation Programs (OMA/PVM Department), Including Appendix D: Guidelines for Collaborative Study Procedures to Validate Characteristics of a Method of Analysis. 2000. Available online: http://www.aoac.org/vmeth/devmethno.htm (accessed on 18 July 2024).
28. Tobiszewski, M.; Namieśnik, J. Greener organic solvents in analytical chemistry. *Curr. Opin. Green Sustain. Chem.* **2017**, *5*, 1–4. [CrossRef]

molecules

MDPI

Article

Systematic Comparison of Extract Clean-Up with Currently Used Sorbents for Dispersive Solid-Phase Extraction

Michelle Peter and Christoph Müller *

Department of Pharmacy, Center for Drug Research, Ludwig-Maximilians-Universität München, 81377 Munich, Germany; michelle.peter@cup.uni-muenchen.de
* Correspondence: christoph.mueller@cup.uni-muenchen.de

Abstract: Dispersive solid-phase extraction (dSPE) is a crucial step for multiresidue analysis used to remove matrix components from extracts. This purification prevents contamination of instrumental equipment and improves method selectivity, sensitivity, and reproducibility. Therefore, a clean-up step is recommended, but an over-purified extract can lead to analyte loss due to adsorption to the sorbent. This study provides a systematic comparison of the advantages and disadvantages of the well-established dSPE sorbents PSA, GCB, and C18 and the novel dSPE sorbents chitin, chitosan, multi-walled carbon nanotube (MWCNT), and Z-Sep® (zirconium-based sorbent). They were tested regarding their clean-up capacity by visual inspection, UV, and GC-MS measurements. The recovery rates of 98 analytes, including pesticides, active pharmaceutical ingredients, and emerging environmental pollutants with a broad range of physicochemical properties, were determined by GC-MS/MS. Experiments were performed with five different matrices, commonly used in food analysis (spinach, orange, avocado, salmon, and bovine liver). Overall, Z-Sep® was the best sorbent regarding clean-up capacity, reducing matrix components to the greatest extent with a median of 50% in UV and GC-MS measurements, while MWCNTs had the largest impact on analyte recovery, with 14 analytes showing recoveries below 70%. PSA showed the best performance overall.

Keywords: dSPE; GC-MS/MS; matrix effects; pesticides; pollutants; QuEChERS

1. Introduction

In 2003, Anastassiades et al. [1] introduced a new sample preparation method for the analysis of pesticides in fruits and vegetables, the QuEChERS sample preparation method (acronym for "quick, easy, cheap, effective, rugged and safe"), which was better adapted to the modern requirements in pesticide analysis than previously common, often over 30-year-old methods [1]. Since then, the QuEChERS concept has been adopted in the EN:15662 [2] and in the AOAC's official method 2007.01 [3] due to its ease of use and broad application. The triumph of this approach has continued for the analysis of environmental contaminants in food and feed matrices such as cabbage and radish [4], wastewater and sludge samples [5], and samples for contamination monitoring in the environment, of animal (e.g., bats [6,7], fish [8], or insects [9]) and plant origin (e.g., beech leaves and spruce needles [10], vine leaves [11], and forage grass [12]), or soil [13].

The QuEChERS method consists of two steps, a salt-assisted liquid–liquid extraction (SALLE), followed by a clean-up step through dispersive solid-phase extraction (dSPE). The use of SALLE is important to achieve rapid phase separation, to salt out proteins, and to maximize the extraction capacity for organic compounds (e.g., pesticides and organic pollutants) [1]. The subsequent clean-up step is necessary to remove a large part of co-extracted organic matrix components, in order to reduce matrix effects [14] and to optimize selectivity and sensitivity [15]. Ideally, this sample preparation grants a high reproducibility in combination with low analyte losses during the procedure. Contrary to previously applied methods, the QuEChERS concept implements dSPE for extract purification. The sorbent is dispersed in the extract during sample clean-up, in contrast to traditional solid-phase extraction (SPE), where the sorbent is packed in plastic cartridges. The original QuEChERS method uses ethylenediamine-*N*-propyl (PSA)-functionalized silica as a sorbent (Figure 1) [1]. PSA has already been used for sample preparation as a sorbent in SPE. Due to its weak anion exchange capacity, it retains polar organic compounds like acids and sugars [16], as well as metal ions through chelation [17].

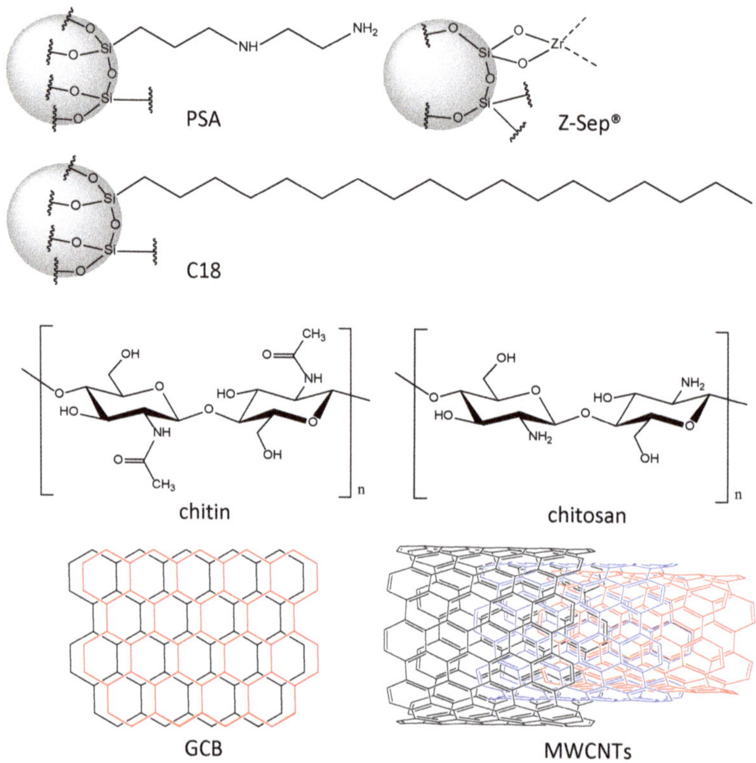

Figure 1. Commonly used dSPE sorbents.

With the alterations in the original QuEChERS method, different sorbents were used for dSPE. The EN:15662 [2] recommends the application of octadecyl-functionalized silica (C18; Figure 1) as well as graphitized carbon black (GCB; Figure 1) as alternative sorbents, depending on the matrix to be analyzed. For fatty matrices such as avocados or olives, the EN:15662 [2] recommends the addition of C18 to dSPE, to remove lipophilic matrix components (e.g., fatty acids). GCB is commonly applied to reduce the amount of large, planar molecules, like pigments (e.g., chlorophyll, carotenoids) and steroids in the matrix extract [18]. The combination of these three sorbents, PSA, C18, and GCB, has proven useful for several complex matrices, providing a solid clean-up with good reproducibility and high recovery rates [6,19–21]. However, the application of GCB for matrix clean-up can negatively influence the recovery of larger, planar analytes [22,23], namely certain pesticides (e.g., thiabendazole; Supplementary Material Figure S1) but also steroid hormones (e.g., testosterone or estradiol; Supplementary Material Figure S1) or polycyclic aromatic hydrocarbons (PAHs; Supplementary Material Figure S1).

Apart from these three sorbents, recent QuEChERS-based methods applied different sorbents that were not part of the regulatory method. Multi-walled carbon nanotubes (MWCNTs) are coaxial closed layers of carbon hexagons with two to fifty nested into one another (Figure 1). They were originally introduced by Sumio Iijiama in 1991 [24] and are mainly used in semiconductor technology. Due to their large surface area and unique absorption properties, MWCNTs are described as a possible alternative to GCB. Both are able to establish π–π interactions and therefore remove nonpolar compounds during sample clean-up [25]. MWCNTs are used for the analysis of pesticides in garlic [26], fruit juices [27], as well as fruit and vegetables [28].

Z-Sep® is a quite recently introduced, commercially available sorbent based on zirconium dioxide-functionalized silica (Figure 1) [29]. According to the supplier, it is suitable for the sample clean-up of fatty matrices when highly lipophilic analytes are of interest, through the removal of lipids and pigments via Lewis acid–base interactions [30]. It has been successfully applied to multiresidue analysis in fish [31], extending the scope of the QuEChERS concept to flame retardants and other emerging environmental pollutants like PAHs.

With growing interest in green chemistry, the applications of biopolymers as sorbents in sample clean-up became popular, as they are available from renewable sources with high prevalence. The glucose derivative chitin, a polymer of *N*-acetylglucosamine (Figure 1), is the second most abundant biopolymer in nature and therefore easily available [32]. It can be isolated through processing the discard of crustaceans like crawfish [33], adding value to crustacean shell waste, since shellfish discard and its deposal pose a challenge for shellfish-producing countries [34]. The use of chitin as a sorbent in QuEChERS-based analytical methods could increase recoveries for pesticides, active pharmaceutical ingredients (APIs), and personal care products (PCPs) in drinking water treatment sludge [35]. Chitin and its deacetylated form chitosan (Figure 1) [34] have been successfully used to adsorb metal ions for the detoxification of water [36]. Due to their molecular structures and functional groups (hydroxyl, amine, or amide), chitin and chitosan are effective for the removal of several organic matrix components, including dyes [37] and lipids [38], similar to PSA.

Moreover, there are many more different sorbents described for their application in sample preparation. Recent developments for solid-phase extraction (SPE) include mesoporous silica which often needs the addition of surface modifiers [39], or sorbents based on magnetic nanoparticles (MNPs) and molecular imprinted polymers (MIPs) [40]. The advancements in sorbent development and availability underline the strong interest in dSPE as a sample clean-up and increase its meaning and scope [40].

However, the different sorbents only get tested for different matrices when they are individually of interest in certain matrices. In this study, we systematically compared seven of the most prominent sorbents in dSPE, PSA, C18, GCB, chitin, chitosan, MWCNTs, and Z-Sep®, regarding their clean-up capacity. We also assessed the recoveries they achieved for a broad range of 98 analytes, including 51 pesticides, 12 active pharmaceutical ingredients (APIs), 12 persistent organic pollutants (POPs), four polycyclic aromatic hydrocarbons (PAHs), 11 personal care product ingredients (PCPs), three flame retardants, two plasticizers, two industrial chemicals, and one indicator for anthropogenic pollution in five different common matrices (spinach, avocado, orange, bovine liver, salmon), to achieve a better understanding of the ideal scope and limitations of each sorbent. For all the chosen matrices, adapted and validated QuEChERS-based analytical methods for pesticide and POP quantification were published [6,22,41–45]. The matrices were chosen to preferably cover a broad scope of applications of the QuEChERS approach. They are of plant (avocado, spinach, and orange) and animal (bovine liver, salmon) origin and cover lipid-dense as well as high-water, high-protein, and/or high-pigment content matrices (Table 1). The different compositions of the chosen matrices pose a variety of challenges to sample clean-up. Spinach extract is especially rich in coloring matrix components, mainly chlorophylls and certain carotenoids and flavonoids, requiring a refined clean-up [41]. Contrary to most plant-based matrices, avocado has a relatively high lipid content (15%), thus posing different challenges to sample preparation [46]. In the avocado, as well as in salmon and liver, the high fat content contributes to the complexity of the matrix, being easily co-extracted during SALLE, thus negatively affecting method performance unless removed from the extract with dSPE [47]. The matrix components in citrus fruits like the orange consist of a variety of flavonoids, coumarins, polyphenols, as well as terpenoids [48].

Furthermore, the citric acid in orange extracts leads to a slightly acidic pH [49], which influences the recovery rate especially of basic analytes. The avocado as well as the orange are part of the most studied fruits in the last years regarding pesticide burden [50]. Liver and fish muscle are especially interesting matrices because lipophilic contaminants (like POPs) can accumulate in the fatty tissue [51].

The tested analytes also covered a broad range of analytes commonly assessed with QuEChERS methods. They included pesticides, which the QuEChERS approach was originally intended for [1], as well as substance classes that were since included in QuEChERS-based methods, such as PAHs [52], PCPs [53] and APIs [54]. These substances are of particular concern in environmental and food analysis. Pesticides and POPs (such as lindane, polychlorinated biphenyls, DDT) are prone to bioaccumulation along the food chain and have been shown to negatively affect the environment, for example by decreasing reproduction, showing endocrine-disruptive effects, and overall increasing mortality rates when found in mammals [55,56]. PAHs, like fluorene or phenanthrene, have been of interest due to their carcinogenic and mutagenic potential [57]. Substances like the UV blocker avobenzone were assessed by the European Chemicals Agency (ECHA) under their community rolling action plan (CoRAP) for reasons like high aggregated tonnage, high exposure, and potential persistency, bioaccumulation, and toxicity (PBT) [58]. APIs, including the opioid codeine [59], the anticonvulsant carbamazepine [60], or the hormone medication 17α-ethinylestradiol [61], have been found in wastewater and drinking water samples. Exposure to APIs can have negative impacts on the environment even in low concentrations, due to the unknown synergistic effects and long-term exposure [59–61]. The UV filter oxybenzone was shown to have genotoxic effects on corals and is a known endocrine disruptor in mammals and fish [62].

With the combination of matrices, analytes, and sorbents tested in this study, we aimed to cover a variety of possible applications of the QuEChERS sample preparation approach and provide a profound overview of which commonly used sorbent would be suitable for different scopes of applications.

Table 1. Distributions of water contents, total lipids (fats), carbohydrates, proteins, and other organic components in the analyzed matrices.

Matrix	Nutrients	Rich in
spinach [63] (mature)	• 92% water • 3% carbohydrates • 3% proteins	• chlorophyll • vitamin A, C, and K • carotenoids and flavonoids
avocado [64] (raw)	• 73% water • 15% total lipids (fats) • 9% carbohydrates	• phytosterols • monounsaturated fatty acids • carotenoids
orange [65] (raw)	• 87% water • 12% carbohydrates	• vitamin C • carotenoids • citric acid
salmon [66] (Atlantic farm-raised, raw)	• 66% water • 20% proteins • 13% total lipids (fats)	• polyunsaturated fatty acids • vitamin B12 and D • cholesterol
liver [67] (pork)	• 71% water • 21% proteins • 4% total lipids (fats)	• cholesterol • bile acids • vitamin A, B12, and D

2. Results and Discussion

2.1. Determination of the Sample Clean-Up Capacity

2.1.1. By Visual Inspection

In the first step, the raw and cleaned-up extracts were visually examined. Figure 2 shows the colors of each extract after clean-up with different dSPE sorbents. The colors were taken from the pictures shown in Supplementary Material Figure S2, for better comparison. The spinach extracts showed no difference in color or intensity except after clean-up with PSA, where they had a lighter, less intense green color. For the orange, avocado, salmon, and liver, the sample clean-up with both GCB and MWCNTs led to a nearly colorless extract, significantly reducing the coloring matrix components, as they are commonly applied to remove large nonpolar molecules like pigments. In the avocado extracts, the application of PSA and Z-Sep® also slightly reduced the intensity of the color of the extract. This effect was also observed for liver extracts, where PSA and Z-Sep®, and additionally chitosan, removed most coloring matrix components. For the salmon extracts, a reduced intensity was also present after the clean-up with Z-Sep®. The colors were extracted from the pictures shown in Supplementary Material Figure S2, where the blank value represents the color of the background. The color transition between Figure 2 and Supplementary Material Figure S2 was achieved by extracting the color of the respective matrix extract from Supplementary Material Figure S2 via the Microsoft Office PowerPoint 365′ eyedropper tool. The colors in Figure 2 have the same RGB code as the according matrix extracts in Supplementary Material Figure S2.

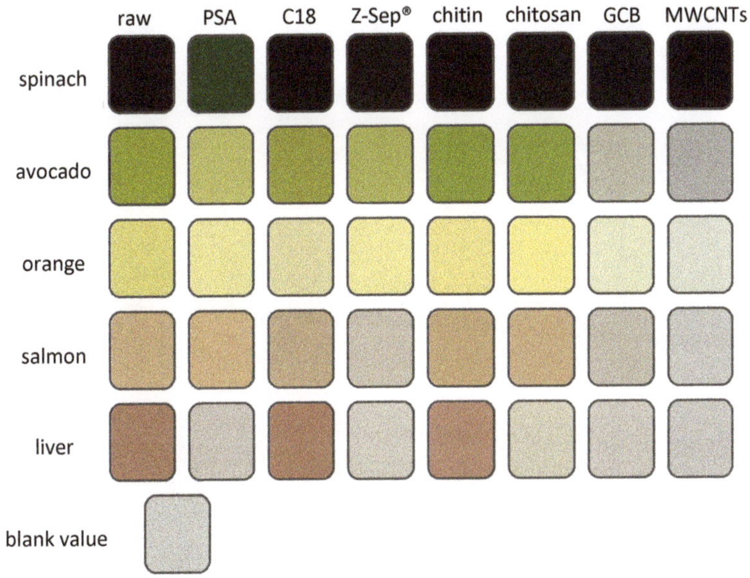

Figure 2. Color intensities from raw extracts compared to clean-up extracts after dSPE. The raw column shows the colors of the extracts before clean-up, and the other columns show the colors of the extracts after clean-up with the respective dSPE sorbents. A reduction in color intensity or change in color tone represents the removal of matrix components (especially of pigments). Colors were extracted from the pictures in Supplementary Material Figure S2, shaped, and rearranged for better comparability.

2.1.2. By UV Measurements

Besides visual inspection, the ability of the seven dSPE sorbents to effectively clean-up raw matrix extracts was determined by UV measurements. As many matrix components absorb at 220 nm, a lower relative extinction corresponded to a cleaner extract, where components absorbing at the specific wavelength were removed. Figure 3 shows the relative extinction of extracts from spinach, avocado, orange, salmon, and liver, each cleaned-up with PSA, C18, Z-Sep®, chitin, chitosan, GCB, or MWCNTs. The distribution of relative extinctions across all matrices (Figure 3A) showed the lowest relative extinction for extracts cleaned-up with PSA and Z-Sep®. This correlated quite well with the observations from the visual inspection, indicating that both PSA and Z-Sep® could remove visible and UV active components from a variety of matrices. However, despite the visible color reduction, the relative extinction after clean-up with GCB and MWCNTs was quite high. The same applied to C18, whereas chitin and chitosan led to moderate relative extinctions. Regarding the detailed distribution of relative extinctions (Figure 3B), it became clear that spinach had the lowest relative extinctions across all sorbents except for Z-Sep®, where avocado yielded the lowest value. This showed that the sorbents were able to remove plenty of matrix components form the spinach extracts, despite nearly no changes in the color intensity of the extracts. The best performing sorbents, regarding the removal of matrix components that absorbed at 220 nm, were PSA for spinach and liver, and Z-Sep® for the avocado, orange, and salmon. C18 removed neither the coloring agents from the extracts nor the matrix components adsorbing at 220 nm. GCB and MWCNTs were not able to remove matrix components absorbing at 220 nm, despite removing most coloring agents from the extracts. This underlines the fact that visual colorless extracts are not necessarily clean in terms of containing interfering matrix components.

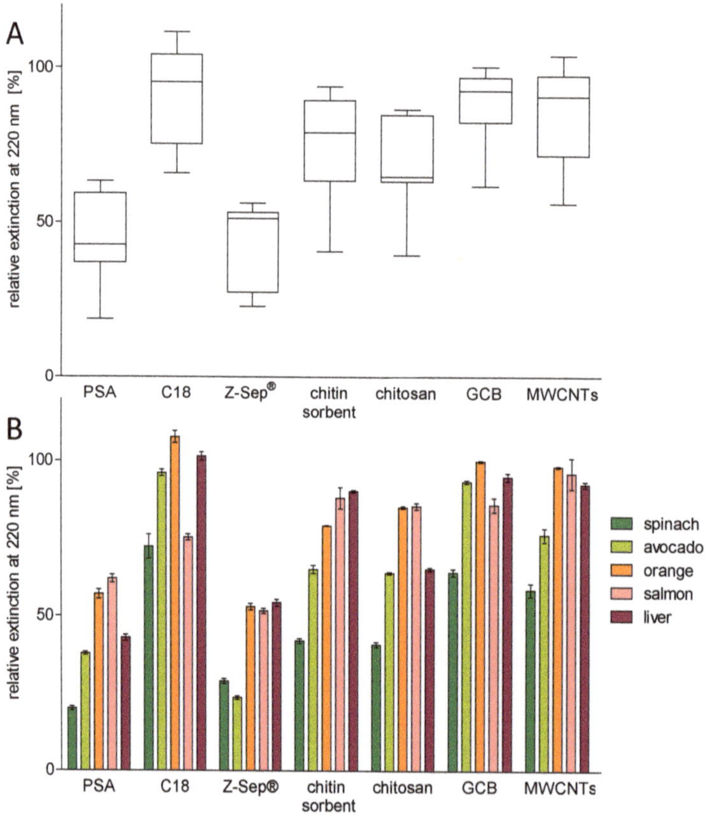

Figure 3. Distribution of the extinction of cleaned-up extracts at 220 nm relative to the extinction of the raw extracts, sorted by the sorbent used for clean-up across all matrices ((**A**), median indicated by the line in the box, whiskers show maximum and minimum values) and for the individual matrices ((**B**), n = 3, error bars show the relative standard deviation).

2.1.3. By MS Measurements

As an additional method to assess the clean-up capacity of the sorbents, the total ion current (TIC) chromatograms of blank and cleaned-up extracts were compared. Supplementary Material Figure S3 shows the TIC chromatograms of each matrix cleaned-up with each sorbent in comparison to the corresponding raw matrix. The peak areas of each matrix cleaned-up with each sorbent are summarized in Supplementary Material Table S1. In Figure 4, the relative TIC peak areas of cleaned-up extracts compared with raw extracts are shown. The distributions of relative peak areas across all matrices (Figure 4A) showed a similar pattern as in the UV measurements (Figure 3A), underlining that PSA and Z-Sep® were able to remove a broad range of matrix components, including gas chromatography-mass spectrometry (GC-MS) detectable components. The other sorbents showed notably higher relative peak areas. The detailed comparison of relative peak areas (Figure 4B) showed that in general, all sorbents were able to remove approximately 50% or more of GC-MS detectable matrix components in the spinach and liver matrices, with C18 reducing the relative peak area for spinach and liver the least. The relative peak areas obtained from the cleaned-up orange extracts also showed quite considerable reduction compared to the raw extracts. Here, PSA, Z-Sep®, chitin, as well as chitosan performed well. The extracts from the salmon matrix only showed a clear reduction in the relative peak area after clean-up with Z-Sep® and PSA. The avocado extracts generally had the lowest reduction in its relative peak areas, with clear effects only after clean-up with PSA, Z-Sep®, and C18. The best performing sorbent regarding the removal of matrix components detectable with GC-MS was Z-Sep® for all matrices.

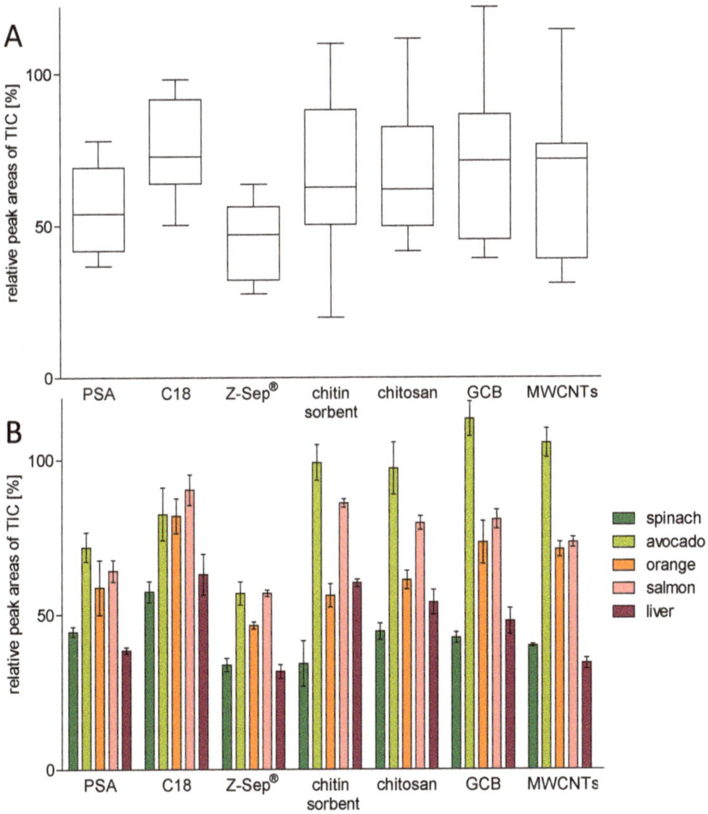

Figure 4. Distribution of mean peak areas in TIC chromatograms (scan m/z 100–600) of cleaned-up extracts relative to raw extracts, sorted by the sorbent used for clean-up across all matrices ((**A**), median indicated by the line in the box, whiskers show maximum and minimum values) and for the individual matrices ((**B**), n = 3, error bars show the relative standard deviation).

On the one hand, the results regarding the GC-MS detectable matrix components correlated quite well with the UV measurements, as in both experiments, PSA and Z-Sep® reduced the relative signals (either peak area or extinction at 220 nm) to the largest extent. C18, chitin, and chitosan showed a moderate clean-up capacity, whereas especially GCB and MWCNTs were not able to reduce matrix components in considerable amounts.

On the other hand, the sorbents did not remove UV-absorbing matrix components as well as they did GC-MS detectable components in the orange and salmon matrices. All sorbents, except PSA and Z-Sep®, reduced the relative peak area to a larger extent than they reduced the relative extinction at 220 nm, indicating that they removed more GC-MS assessable matrix components than UV-absorbing components. The relative peak areas obtained from the avocado extract could not be lowered through clean-up with chitin, chitosan, GCB, and MWCNTs.

2.2. Determination of the Analyte Recovery

Besides the removal of interfering matrix components, a crucial quality characteristic was the recovery of analytes. Ideally, the sample preparation would not remove analytes of interest while still providing a relatively clean extract. Guidelines for method validation like SANTE/11312 [68] or CXG 90-2017 (Guidelines on Performance Criteria for Methods of Analysis for the Determination of Pesticide Residues in Food and Feed) [69] define the parameters to evaluate this effect as recovery. For an analytical method to meet validation criteria, the recovery must lie within a range of 70 to 120% (SANTE/11312 [68] and CXG 90-2017 [69]). In this study, the influence of PSA, C18, Z-Sep®, chitin, chitosan, GCB, and MWCNTs on the recovery of the 98 analytes (Supplementary Material Table S3) with different physicochemical properties was tested to gain a better understanding of their ideal applications in sample preparation depending on the analytes of interest and sample matrix. Figure 5 shows the distribution of recoveries of all analytes across all five matrices obtained after sample clean-up with different dSPE sorbents. It is striking that while most sorbents achieved a median recovery well around 100% (from 94% for PSA to 104% for chitin), the application of MWCNTs led to clearly lower recoveries (median 86%) than the rest of the sorbents. However, when calculating the mean across all analytes, every sorbent for every matrix laid within the required range of 70 to 120% (Figure 5, upper left).

A more detailed analysis showed the influence of the different sorbents on the analytes spiked into the individual matrices (Figure 5). For all sorbents, the mean recovery as well as the box, which equaled the middle 50% of the data, laid within the recommended range of 70 to 120%. In this study, special interest was paid to analytes with recoveries below 70%, as the sorbents were especially accountable for these low values. A list of all analytes and their respective recoveries is provided in Supplementary Material Table S2.

After clean-up with chitosan, no analyte had a recovery below 70% and only very few analytes reached recoveries above 120%. Chitosan therefore did not negatively affect the analyte recoveries in the five tested matrices, so it is a suitable sorbent for a broad range of applications.

Its structural derivative, chitin, showed a similar performance only with one analyte (codeine, 68%; Supplementary Material Table S2) below 70% recovery in the spinach extract. However, as a recovery of 68% was only slightly under the criterium, chitin could nonetheless be a suitable sorbent for the analysis of codeine and structurally familiar components.

Another sorbent that led to mostly recommended recoveries with a low scatter around the median was PSA. Only in the spinach, salmon and liver matrices did some analytes show recoveries below 70% (clotrimazole 66% in spinach, lilial 52% in spinach, 62% in salmon, and 68% in liver, and avobenzone 67% in salmon (structures of the analytes are given in Supplementary Material Figure S1)). As a weak anion exchange agent, PSA is commonly used to remove polar organic matrix compounds, such as acids and sugars [16]. Both lilial and avobenzone are susceptible to interactions with PSA. Lilial as an aldehyde could potentially form a covalent imine bond with the amine groups in PSA [70].

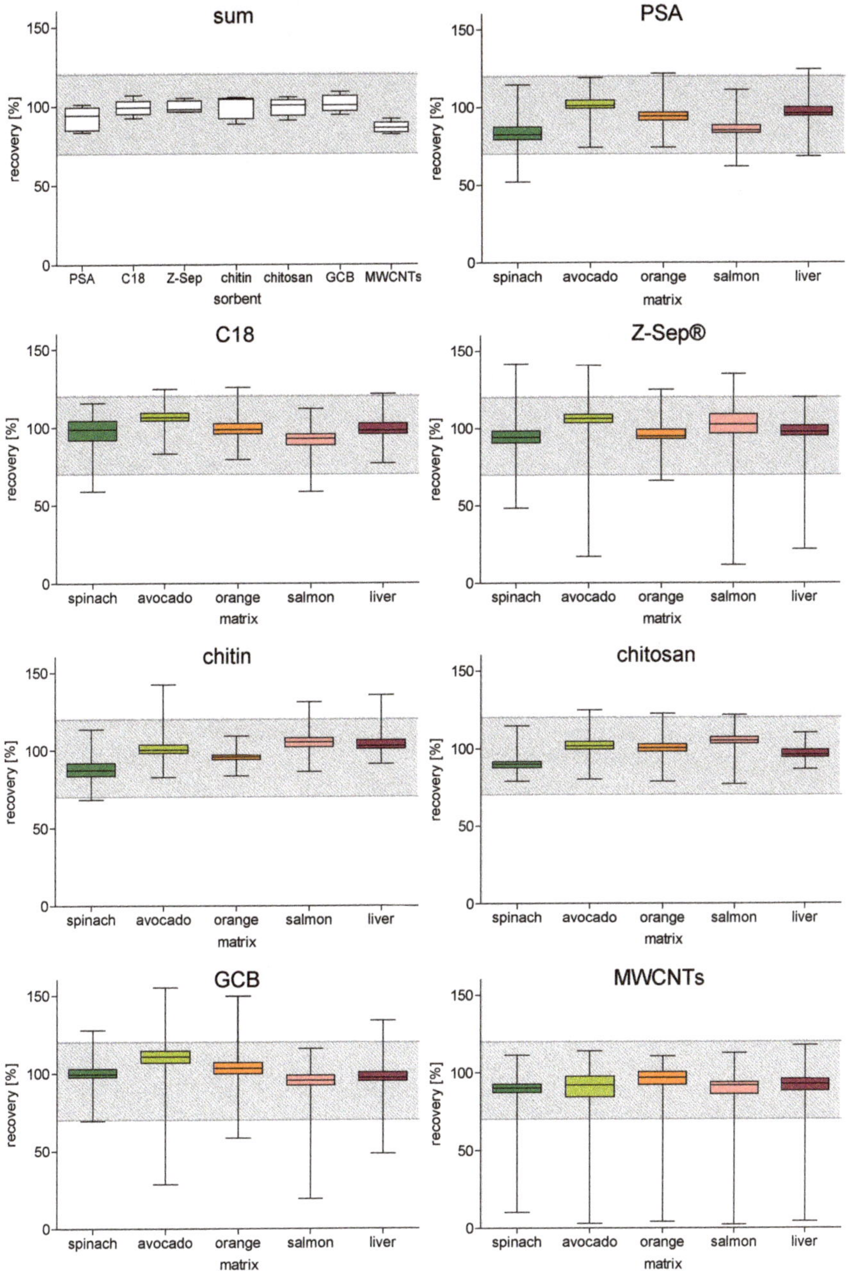

Figure 5. Analysis of the recovery data. For each matrix cleaned-up with each sorbent, the mean recovery across all analytes was calculated, resulting in one value per matrix and sorbent. The box plots represent the recoveries for the five different matrices, the lines represent the median recoveries, and the whiskers show the minimum and maximum values. The table containing all individual recovery values can be found in Supplementary Material Table S2.

Avobenzone in general showed rather low recoveries after sample clean-up with PSA (from 75% in spinach to 90% in orange). The distributions of recoveries for C18 were quite similar to PSA; however, there were different analytes that did not comply with the recovery criteria. In the spinach extract cleaned-up with C18, the outlier was the herbicide isoproturon, with a recovery of 59%, and in the salmon extract, the herbicide atrazine had a recovery of 59%.

After clean-up with Z-Sep®, more analytes showed recoveries below 70%, and the outliers were further away from the recommended recovery. Avobenzone had recoveries from 12% in salmon extract to 48% in spinach extract, and only in the orange extract was 73% reached. Through its tautomerism, the 1,3 diketone avobenzone is an ideal bidentate ligand for the zirconium in Z-Sep®, which adsorbs avobenzone to a greater extent than PSA. The other outliers included benzophenone-1 (68% in salmon, 69% in avocado), codeine (65% in spinach, 66% in orange, 68% in salmon), imidacloprid (63% in avocado), and oxybenzone (68% in salmon). Remarkably, besides avobenzone, the two ketones benzophenone-1 and oxybenzone had low recoveries, as the oxygen of the ketol group (hydroxy ketone group) can act as a Lewis base and likely interacted with Z-Sep®, forming a Lewis acid–base adduct. Interestingly, PSA only affected the recovery of avobenzone, but not of other ketones (benzophenone-1 or oxybenzone), while Z-Sep® had a lesser influence on the recovery of lilial. The interaction between ketones and Z-Sep® however seems to have been weaker than between the β-diketone and Z-Sep®, as their recoveries were still close to 70%.

The sorbent GCB led to recoveries below 70% for avobenzone, phenanthrene, pyrene, and tetraconazole. Besides avobenzone, which seemed like a difficult analyte for most sorbents, the analytes affected by GCB were planar molecules with multiple aromatic rings that could easily form π–π interactions with the sp^2-hybridized carbons of the graphene structure [71]. Pyrene had by far the lowest recoveries (from 19% in salmon to 58% in orange), and only in spinach did the recovery of 77% comply with the requirements. This might be due to the large amounts of pigments present in the spinach extract, potentially masking more active sites of the sorbent. This highlights the impact GCB can have on planar analytes during sample clean-up.

However, the lowest recoveries were clearly obtained when MWCNTs were used as sorbent for dSPE. A total of 14 analytes had recoveries below 70% in at least one of the matrices after clean-up with MWCNTs. Like with GCB, the large planar analytes showed the lowest recoveries, but the offset was significantly higher. Pyrene for example was not even detectable in the avocado, salmon, and liver extracts after clean-up with MWCNTs, and it had recoveries of 4% in the orange and 10% in the spinach extracts. Other planar analytes, such as fluorene, phenanthrene, and thiabendazole, were severely affected as well. Furthermore, avobenzone, benzophenone-1, and oxybenzone also had low recoveries, but since they were also affected by other sorbents, this might be due to other factors of the clean-up procedure, possibly the slightly acidic pH of the citrate-buffered extraction system [3], which was not perfectly suitable for these analytes. Regarding the severe impact of MWCNTs on a large number of analytes in comparison to GCB, it became clear that GCB was the better option for the removal of large, planar matrix components. Both sorbents performed equally well in reducing the amounts of pigments from colored extracts and possibly other planar matrix components, but GCB did not impact the recovery of planar analytes of interest to the same extent as MWCNTs did.

3. Materials and Methods

3.1. Materials

3.1.1. Chemicals

All analytes and the deuterated internal standards (azoxystrobin-d_4, chlorpyrifos-d_4, and pyrene-d_{10}) had a purity >98% and were purchased either from BLDpharm (Kaiserslautern, Germany), ChemPUR (Karlsruhe, Germany), EDQM (Strasbourg, France), HPC Standards GmbH (Cunnersdorf, Germany), Merck (Darmstadt, Germany), or Tokyo Chemical Industries (TCI, Tokyo, Japan). A complete list of all 98 analytes is provided in Supplementary Material

Table S3. Dispersive SPE bulk sorbents of ethylenediamine-*N*-propyl-functionalized silica (PSA), octadecyl-functionalized silica (C18), and graphitized carbon black (GCB) were obtained from Agilent Technologies (Santa Clara, CA, USA). The bulk of the multi-walled carbon nanotubes (MWCNTs, outside diameter 20–30 nm) was purchased from abcr (Karlsruhe, Germany). Chitin and chitosan (unknown purity) were obtained from TCI (Tokyo, Japan). Z-Sep® was bought from Merck (Sant Louis, MO, USA). Acetonitrile (ACN) and isopropanol of HPLC grade were purchased from VWR (Darmstadt, Germany). 3-Ethoxy-1,2-propanediol (98%) was obtained from BLDpharm (Kaiserslautern, Germany), and L-gluconic acid γ-lactone (95%), D-sorbitol (99%), sodium citrate dihydrate ($\geq$99%), and disodium hydrogen citrate sesquihydrate ($\geq$99%) were purchased from Merck (Darmstadt, Germany). Shikimic acid ($\geq$98%) was obtained from Carl Roth (Karlsruhe, Germany). Anhydrous magnesium sulphate ($MgSO_4$, $\geq$98%) was purchased from Grüssing (Filsum, Germany). Sodium chloride (NaCl, p.a.) was obtained from Bernd Kraft (Duisburg, Germany).

3.1.2. Matrices

For the preparation of blank matrices, the spinach, orange, avocado, salmon and bovine liver were bought at a local grocery store. Each matrix was homogenized with an Ultra Turrax® T25 basic (IKA Labortechnik, Staufen, Germany) at 13,000 rpm for 15 min. After each matrix, the rotor was cleaned for 15 min with ultrapure water and isopropanol. After homogenization, the matrices were weighed into 50 mL centrifuge tubes in aliquots of 10.0 g ($\pm$0.5 g) (spinach and orange, according to EN:15662 [2]) or 5.0 g ($\pm$0.1 g) for the avocado [2], salmon, and liver [6], respectively. Where applicable, the amount of matrix per aliquot was corresponding to the EN:15662 [2]. For the liver, 5.0 g was used according to Schanzer et al. [6], and an equal amount was used for the salmon to ensure comparability. The homogenized matrix aliquots were stored in a freezer at –20 °C.

3.2. *Reagents*

3.2.1. Stock Solutions

Stock solutions with a concentration of 1 mg mL^{-1} were prepared for each analyte and for the two internal standards azoxystrobin-d_4 and chlorpyrifos-d_{10}. The stock solutions were combined into five working solutions, containing analytes grouped by their chemical properties (alcohols, acids, nonpolar substances, amines/amides, and ethers/esters), with a concentration of 10 µg mL^{-1} each [72]. An internal standard working solution was prepared accordingly. The stock and working solutions were stored at –20 °C and allowed to settle to room temperature for 1 h before use.

3.2.2. Analyte Protectant Solution

The analyte protectant solution (AP) was prepared from 3-ethoxy-1,2-propanediol (200 mg mL^{-1}), L-gluconic acid γ-lactone (10 mg mL^{-1}), sorbitol (5 mg mL^{-1}), and shikimic acid (5 mg mL^{-1}) in ACN and water (6:4, (v/v)) according to an application note from the EU Reference Laboratories for Residues of Pesticides (EURL-SRM) from 2013 [73].

3.2.3. Buffer Salt Mixture

The buffer salt mixture for the salting out step of the QuEChERS sample preparation consisted of anhydrous $MgSO_4$, NaCl, sodium citrate dihydrate, and sodium hydrogen citrate sesquihydrate in a ratio of 8:2:2:1 according to EN:15662 [2]. Therefore, 4.00 g ($\pm$0.20 g) anhydrous $MgSO_4$, 1.00 ($\pm$0.05 g) NaCl, 1.00 ($\pm$0.05 g) sodium citrate dihydrate, and 0.50 g ($\pm$0.03 g) disodium citrate dihydrate were weighed into 15 mL centrifuge tubes.

3.2.4. dSPE Mixture

For the preparation of the dSPE mixtures, anhydrous $MgSO_4$ and each respective sorbent were combined in 15 mL centrifuge tubes and vigorously shaken by hand for 1 min, except for the mixture containing GCB, which was mixed with anhydrous $MgSO_4$ with a mortar and pestle. The ratio of anhydrous $MgSO_4$ to sorbent was 3:1 (for PSA, C18,

Z-Sep®, chitin, and chitosan) and 20:1 for GCB and MWCNTs [2]. Chitin and chitosan were washed/cleaned with *tert*-butyl methyl ether (4 times) and ACN (4 times) and evaporated to dryness with a rotary evaporator. Detailed information on the amounts of sorbent and anhydrous $MgSO_4$ used per mL extract for clean-up is given in Table 2.

Table 2. Compositions of dSPE sorbent mixtures.

Sorbent	Amount of Sorbent [mg]	Amount of $MgSO_4$ [mg]
PSA	50	150
C18	50	150
Z-Sep®	50	150
chitin	50	150
chitosan	50	150
GCB	7.5	150
MWCNTs	7.5	150

3.3. *Instrumentation*

3.3.1. Sample Preparation Equipment

For sample homogenization, an Ultra Turrax® T25 basic (IKA Labortechnik, Staufen, Germany) was used. For the centrifugation of the 50 mL centrifuge tubes, a Heraeus Megafuge (VWR, Darmstadt, Germany) was used. Centrifugation of the 15 mL centrifuge tubes was performed with an EBA 20 centrifuge (Hettich, Tuttlingen, Germany) and for 2 mL microcentrifuge tubes, a 5415 D centrifuge from Eppendorf (Hamburg, Germany) was used. Shaking steps were performed with a Vortex Genie (Scientific Industries, Bohemia, NY, USA).

3.3.2. UV–Vis Spectrophotometer

The clean-up capacity of the sorbents was assessed with UV measurements at 220 nm, performed with an M2e plate reader from Molecular Devices (San Jose, CA, USA).

3.3.3. Gas Chromatography-Tandem Mass Spectrometry (GC-MS/MS)

Analysis was performed with a 7890 gas chromatograph coupled with a 7010B triple quadrupole mass spectrometer, which was equipped with a high-efficiency ion source (HES) (all from Agilent, Santa Clara, CA, USA). For the injection of 1 μL, a PAL3 RSI autosampler from CTC analytics (Zwingen, Switzerland) was used. The columns for separation were two Agilent J&W HP 5MS Ultra Inert capillary columns (15 m × 250 μm × 0.25 μm) coupled with a capillary flow technology (CFT) backflush device. The GC was equipped with a multi-mode inlet (MMI), operated in solvent vent mode. All parameters of the GC-MS/MS method were applied as in our previously published QuEChERS approach [72]. A summary of the multiple reaction monitoring (MRM) conditions for all 98 analytes is listed in Supplementary Material Table S3.

3.4. *Methods*

All experiments were performed for all seven tested dSPE sorbents with the five matrices for spinach, avocado, orange, liver, and salmon, in triplicates.

3.4.1. QuEChERS Sample Preparation

The sample preparation of the plant-based matrices of spinach, avocado, and orange, was performed according to the EN:15662 [2]. As the sample preparation of matrices of animal origin, like salmon and liver, is not specified in the EN:15662 [2], the clean-up step was performed according to the procedure for the analysis of pesticides and persistent organic pollutants (POPs) in animal liver tissues described by Schanzer et al. [6]. A defined portion (Table 3) of the homogenized matrix was weighed into a 50 mL centrifuge tube and an aliquot of ultrapure water was added (Table 3). Then, the samples were shaken with a Vortex Genie for 30 s, and 10 mL ACN were added before another shaking step of

30 s, followed by 15 min of extraction time. Afterwards, the buffer salt mixture was added, and the samples were instantly shaken by hand to prevent the formation of agglomerates. The samples were then again shaken with a Vortex Genie and centrifuged for 5 min at 3600× *g*. The organic upper layer (raw extract) was used for further clean-up with different dSPE sorbents (Table 2). Therefore, an aliquot of 1 mL extract was transferred to a 2 mL microcentrifuge tube containing 200 mg or 157.5 mg of dSPE mixture as described in Chapter 3.2.4. The sample was again shaken by hand followed by 30 s shaking with a Vortex Genie and then centrifuged for 5 min at 12,000× *g*. Aliquots of the supernatants were taken for further analysis. For the determination of analyte recovery, blank extracts (every matrix cleaned-up with every sorbent) were spiked one time before and one time after clean-up with 50 µL (before) or 5 µL (after) of a standard mix containing all analytes at 10 µg mL^{-1} to obtain final nominal concentrations of 50 ng mL^{-1}. Extracts for other experiments were not spiked with analytes.

Table 3. Ratios between added water [mL] per g sample matrix in relation to the sample matrix.

Matrix	Homogenized Matrix [g]	Added Volume H_2O [mL]
spinach	10.00	0
orange	10.00	0
avocado	5.00	6
liver	5.00	5
salmon	5.00	5

3.4.2. Determination of the Sample Clean-Up Capacity

By Visual Inspection

After sample preparation and sample clean-up, the colors of the final extracts from each sorbent and matrix were visually examined.

By UV Measurements

For the assessment of the clean-up capacity of the different sorbents, the extinction at 220 nm wavelength of the cleaned-up extracts was compared to the raw extracts taken from the sample preparation before the dSPE step. Pure ACN was measured as a blank. Three hundred microliters of the solvent, raw extracts, and cleaned-up extracts were transferred to a 96-well fused silica plate and extinction at 220 nm was measured. The relative extinction was calculated according to Equation (1):

$$relative\ extinction\ [\%] = \frac{extinction\ (clean_{220nm})}{extinction\ (raw_{220nm})} \times 100 \quad (1)$$

By MS Measurements

Additionally, extracts of each matrix cleaned-up with each sorbent were analyzed with GC-MS/MS used as described in Chapter 3.3.3. Instead of the MRM mode, the scan mode from *m/z* 100 to 600 was used and the total ion current (TIC) was summed up for every extract. The peak area was determined by using Agilent's Agile2 integration algorithm in the Qualitative Analysis Navigator program (version B.08.00 from 2016).

3.4.3. Determination of the Analyte Recovery

To determine the influence of different sorbents on analyte recovery, the extract of each matrix was spiked before and after the sample clean-up with each of the seven sorbents (PSA, C18, Z-Sep®, chitin, chitosan, GCB, and MWCNTs). The samples were measured with GC-MS/MS and recovery was calculated according to Equation (2):

$$recovery = \frac{area\ (after)}{area\ (before)} \times 100 \quad (2)$$

4. Conclusions

The clean-up capacity of each sorbent (PSA, C18, Z-Sep®, chitin, chitosan, GCB, and MWCNTs) was assessed in three different ways: (1) by evaluating the color intensity of the cleaned-up extracts; (2) by measuring the relative extinction at 220 nm; and (3) by comparing the summed TIC peak areas. Considering all three parameters, Z-Sep® showed the greatest clean-up capacity of the tested sorbents, reducing the GC-MS detectable matrix components in all five matrices to the largest extent. It also performed well in the UV experiment, removing most of the matrix components in the avocado, orange, and salmon extracts. This showed that Z-Sep® was able to remove a broad range of different matrix components. Regarding analyte recovery, Z-Sep® showed a negative influence on many analytes, second only to MWCNTs. The exact identification of the removed matrix components was not possible with the experimental setup. However, the aim of this study was to assess the capacities of different sorbents to reduce co-extracted matrix components while not negatively affecting analyte recovery. PSA provided the second-best result regarding clean-up capacity, while only negatively affecting the recovery of fewer, particular analytes, resulting in overall the best performance of all tested sorbents. The sorbents with a medium clean-up capacity (chitin and chitosan) especially had the least negative influence on the recovery rates of all tested analytes. GCB and MWCNTs only performed well in the visual inspection of color intensities, with contrary results in the other clean-up capacity experiments. Considering analyte recovery, GCB also showed mediocre results. Large planar molecules such as PAHs (see Supplementary Material Figure S1) showed low recoveries when GCB or MWCNTs were used. In general, analytes with one or more (hydroxyl-) carbonyl groups, such as lilial or avobenzone (see Supplementary Material Figure S1) were particularly susceptible to recovery loss. Overall, this study showed that the choice of the ideal sorbent is strongly dependent on the purpose of the applied method (Figure 6). If analytes that are difficult to access are of interest, the use of Z-Sep® and GCB might not be sufficient, as they can lead to very low recoveries. Instead, chitin or chitosan could be viable options, providing a good compromise between clean extracts and few negative effects on analytes. If analyte recovery is not the crucial point, but clean-up is more important, Z-Sep® is the best choice. PSA, as the most established dSPE sorbent, is still the best choice for most QuEChERS applications. The implementation of MWCNTs to QuEChERS-based sample preparation does not seem beneficial in any case. For complex matrices and difficult analytical purposes, individual consideration and a combination of different sorbents is necessary and cannot be replaced by this study. However, the published data provide a valuable starting point for future QuEChERS method development.

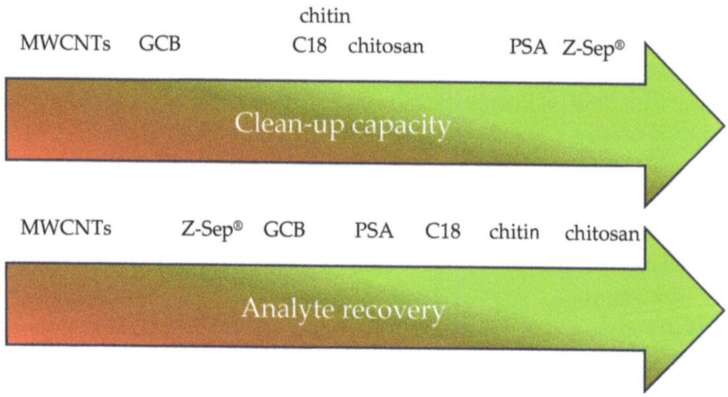

Figure 6. Rankings of seven tested sorbents regarding their performance in clean-up capacity experiments and affecting analyte recovery.

Supplementary Materials: The following supporting information can be downloaded at: https://www.mdpi.com/article/10.3390/molecules29194656/s1, Figure S1: Structures of analytes mentioned in the text; Figure S2: Colors of extracts of spinach (A), avocado (B), orange (C), salmon (D), and liver (E) matrices after clean-up with seven different dSPE sorbents; Figure S3: TIC chromatograms of GC-MS scan data (*m/z* 100-600) of raw (black) and cleaned-up (red) matrix extracts.Table S1: Peak areas of TIC (*m/z* 100-600) of matrix extracts cleaned-up with seven different sorbents (PSA, C18, Z-Sep®, chitin, chitosan, GCB and MWCNTs. The peak area was determined by using Agilent's Agile2 integration algorithm in the Qualitative Analysis Navigator program (version B.08.00 from 2016); n = 3, mean given in counts (absolute) or % (relative)); Table S2: List of all analytes with their recoveries in the respective matrices (n = 3, mean given in %). Recoveries below 70% are indicated in *red*; Table S3: List of all analytes, including classification, retention time (RT), applied MRM transitions (quantifier in bold letters), and collision energy (CE). API, active pharmaceutical ingredient; PAH, polycyclic aromatic hydrocarbon; POP, persistent organic pollutant. For analytes with two or three isomers, all retention times are given.

Author Contributions: Conceptualization, M.P. and C.M.; methodology, M.P and C.M.; software, M.P.; formal analysis, M.P. and C.M.; investigation, M.P.; resources, C.M.; data curation, M.P. and C.M.; writing—original draft preparation, M.P. and C.M.; writing—review and editing, M.P. and C.M.; visualization, M.P.; supervision, C.M.; project administration, C.M. All authors have read and agreed to the published version of the manuscript.

Funding: This research received no external funding.

Data Availability Statement: Data are available from the corresponding author.

Acknowledgments: M.P. and C.M. thank Franz Bracher for providing his laboratories, resources, and discussions. We also thank Carolin Bauernschmidt for performing the UV measurements.

Conflicts of Interest: The authors declare no conflicts of interest.

References

1. Anastassiades, M.; Lehotay, S.J.; Štajnbaher, D.; Schenck, F.J. Fast and easy multiresidue method employing acetonitrile extraction/partitioning and "dispersive solid-phase extraction" for the determination of pesticide residues in produce. *J. AOAC Int.* **2003**, *86*, 412–431. [CrossRef] [PubMed]
2. *EN 15662:2018-07*; Foods of Plant Origin—Multimethod for the Determination of Pesticide Residues Using GC- and LC-Based Analysis following Acetonitrile Extraction/Partitioning and Clean-Up by Dispersive SPE—Modular QuEChERS-Method. European Comittee for Standardization: Brussels, Belgium, 2018.
3. Lehotay, S.; Neil, M.; Tully, J.; Valverde, A.; Contreras, M.; Mol, H.; Heinke, V.; Anspach, T.; Lach, G.; Fussell, R.; et al. Determination of pesticide residues in foods by acetonitrile extraction and partitioning with magnesium sulfate: Collaborative study. *J. AOAC Int.* **2007**, *90*, 485–520. [CrossRef] [PubMed]
4. Nguyen, T.D.; Yu, J.E.; Lee, D.M.; Lee, G.H. A multiresidue method for the determination of 107 pesticides in cabbage and radish using QuEChERS sample preparation method and gas chromatography mass spectrometry. *Food Chem.* **2008**, *110*, 207–213. [CrossRef] [PubMed]
5. Cavaillé, L.; Kim, C.; Bounouba, M.; Zind, H.; Claparols, C.; Riboul, D.; Pinelli, E.; Albasi, C.; Bessiere, Y. Development and validation of QuEChERS-based extraction for quantification of nine micropollutants in wastewater treatment plant. *Anal. Bioanal. Chem.* **2021**, *413*, 5201–5213. [CrossRef] [PubMed]
6. Schanzer, S.; Kröner, E.; Wibbelt, G.; Koch, M.; Kiefer, A.; Bracher, F.; Müller, C. Miniaturized multiresidue method for the analysis of pesticides and persistent organic pollutants in non-target wildlife animal liver tissues using GC-MS/MS. *Chemosphere* **2021**, *279*, 130434. [CrossRef]
7. Kuzukiran, O.; Simsek, I.; Yorulmaz, T.; Yurdakok-Dikmen, B.; Ozkan, O.; Filazi, A. Multiresidues of environmental contaminants in bats from turkey. *Chemosphere* **2021**, *282*, 131022. [CrossRef]
8. Molina-Ruiz, J.M.; Cieslik, E.; Cieslik, I.; Walkowska, I. Determination of pesticide residues in fish tissues by modified QuEChERS method and dual-d-SPE clean-up coupled to gas chromatography–mass spectrometry. *Environ. Sci. Pollut. Res.* **2015**, *22*, 369–378. [CrossRef]
9. Stöckelhuber, M.; Müller, C.; Vetter, F.; Mingo, V.; Lötters, S.; Wagner, N.; Bracher, F. Determination of pesticides adsorbed on arthropods and gastropods by a micro-QuEChERS approach and GC–MS/MS. *Chromatographia* **2017**, *80*, 825–829. [CrossRef]
10. Löbbert, A.; Schanzer, S.; Krehenwinkel, H.; Bracher, F.; Müller, C. Determination of multi pesticide residues in leaf and needle samples using a modified QuEChERS approach and gas chromatography-tandem mass spectrometry. *Anal. Methods* **2021**, *13*, 1138–1146. [CrossRef]
11. Keklik, M.; Golge, O.; González-Curbelo, M.Á.; Kabak, B. Determination of pesticide residues in vine leaves using the QuEChERS method and liquid chromatography-tandem mass spectrometry. *Foods* **2024**, *13*, 909. [CrossRef]

12. Wu, X.; Li, T.; Feng, H.; Xie, Y.; Liu, F.; Tong, K.; Fan, C.; Liu, Y.; Chen, H. Multi-residue analysis of 206 pesticides in grass forage by the one-step quick, easy, cheap, effective, rugged, and safe method combined with ultrahigh-performance liquid chromatography quadrupole orbitrap mass spectrometry. *J. Sep. Sci.* **2022**, *45*, 2520–2528. [CrossRef] [PubMed]
13. González-Curbelo, M.Á.; Varela-Martínez, D.A.; Riaño-Herrera, D.A. Pesticide-residue analysis in soils by the QuEChERS method: A review. *Molecules* **2022**, *27*, 4323. [CrossRef] [PubMed]
14. Ryska, M. How to deal with the "matrix effect" as an unavoidable phenomenon. *Eur. J. Mass Spectrom.* **2015**, *21*, 423–432. [CrossRef] [PubMed]
15. Mona, S.; Sayyed Hossein, H.; Massoud, K. Modern sample preparation techniques: A brief introduction. In *Sample Preparation Techniques for Chemical Analysis*; Massoud, K., Ed.; IntechOpen: Rijeka, Croatia, 2021; Ch. 2.
16. Schenck, F.J.; Lehotay, S.J.; Vega, V. Comparison of solid-phase extraction sorbents for cleanup in pesticide residue analysis of fresh fruits and vegetables. *J. Sep. Sci.* **2002**, *25*, 883–890. [CrossRef]
17. Oshita, D.; Jardim, I.C.S.F. Comparison of different sorbents in the QuEChERS method for the determination of pesticide residues in strawberries by LC–MS/MS. *Chromatographia* **2014**, *77*, 1291–1298. [CrossRef]
18. Islam, A.K.M.M.; Lee, H.-S.; Ro, J.-H.; Kim, D.; Kwon, H. Application of high-surface-area graphitized carbon black with primary secondary amine as an alternative quick, easy, cheap, effective, rugged, and safe cleanup material for pesticide multi-residue analysis in spinach. *J. Sep. Sci.* **2019**, *42*, 2379–2389. [CrossRef]
19. Walorczyk, S. Validation and use of a QuEChERS-based gas chromatographic–tandem mass spectrometric method for multiresidue pesticide analysis in blackcurrants including studies of matrix effects and estimation of measurement uncertainty. *Talanta* **2014**, *120*, 106–113. [CrossRef]
20. Cunha, S.C.; Lehotay, S.J.; Mastovska, K.; Fernandes, J.O.; Beatriz, M.; Oliveira, P.P. Evaluation of the QuEChERS sample preparation approach for the analysis of pesticide residues in olives. *J. Sep. Sci.* **2007**, *30*, 620–632. [CrossRef]
21. Lehotay, S.J.; Son, K.A.; Kwon, H.; Koesukwiwat, U.; Fu, W.; Mastovska, K.; Hoh, E.; Leepipatpiboon, N. Comparison of QuEChERS sample preparation methods for the analysis of pesticide residues in fruits and vegetables. *J. Chromatogr. A* **2010**, *1217*, 2548–2560. [CrossRef]
22. Zhao, L.; Stevens, J. Optimizing Recoveries of Planar Pesticides in Spinach Using Toluene and Agilent Bond Elut QuEChERS AOAC Kit with Graphitized Carbon. Available online: https://www.agilent.com/Library/applications/5990-4247EN.pdf (accessed on 22 August 2024).
23. Wilkowska, A.; Biziuk, M. Determination of pesticide residues in food matrices using the QuEChERS methodology. *Food Chem.* **2011**, *125*, 803–812. [CrossRef]
24. Iijima, S. Helical microtubules of graphitic carbon. *Nature* **1991**, *354*, 56–58. [CrossRef]
25. Liang, X.; Liu, S.; Wang, S.; Guo, Y.; Jiang, S. Carbon-based sorbents: Carbon nanotubes. *J. Chromatogr. A* **2014**, *1357*, 53–67. [CrossRef] [PubMed]
26. Du, D.; Wang, M.; Zhang, J.; Cai, J.; Tu, H.; Zhang, A. Application of multiwalled carbon nanotubes for solid-phase extraction of organophosphate pesticide. *Electrochem. Commun.* **2008**, *10*, 85–89. [CrossRef]
27. Ravelo-Pérez, L.M.; Hernández-Borges, J.; Rodríguez-Delgado, M.A. Multi-walled carbon nanotubes as efficient solid-phase extraction materials of organophosphorus pesticides from apple, grape, orange and pineapple fruit juices. *J. Chromatogr. A* **2008**, *1211*, 33–42. [CrossRef] [PubMed]
28. Zhao, P.; Wang, L.; Zhou, L.; Zhang, F.; Kang, S.; Pan, C. Multi-walled carbon nanotubes as alternative reversed-dispersive solid phase extraction materials in pesticide multi-residue analysis with QuEChERS method. *J. Chromatogr. A* **2012**, *1225*, 17–25. [CrossRef]
29. Tuzimski, T.; Szubartowski, S. Method development for selected bisphenols analysis in sweetened condensed milk from a can and breast milk samples by HPLC–DAD and HPLC-QqQ-MS: Comparison of sorbents (Z-Sep, Z-Sep plus, PSA, C18, chitin and EMR-Lipid) for clean-up of QuEChERS extract. *Molecules* **2019**, *24*, 2093. [CrossRef]
30. Sigma-Aldrich. Analysis of Compounds in Fatty Matrices. Available online: https://www.sigmaaldrich.com/deepweb/assets/sigmaaldrich/marketing/global/documents/119/950/12023_T414114_Compounds_in_Fatty_Matrices.pdf (accessed on 1 March 2024).
31. Sapozhnikova, Y.; Lehotay, S.J. Multi-class, multi-residue analysis of pesticides, polychlorinated biphenyls, polycyclic aromatic hydrocarbons, polybrominated diphenyl ethers and novel flame retardants in fish using fast, low-pressure gas chromatography–tandem mass spectrometry. *Anal. Chim. Acta* **2013**, *758*, 80–92. [CrossRef]
32. Tharanathan, R.N.; Kittur, F.S. Chitin—The undisputed biomolecule of great potential. *Crit. Rev. Food Sci. Nutr.* **2003**, *43*, 61–87. [CrossRef]
33. No, H.K.; Meyers, S.P.; Lee, K.S. Isolation and characterization of chitin from crawfish shell waste. *J. Agric. Food Chem.* **1989**, *37*, 575–579. [CrossRef]
34. Shahidi, F.; Arachchi, J.K.V.; Jeon, Y.-J. Food applications of chitin and chitosans. *Trends Food Sci.* **1999**, *10*, 37–51. [CrossRef]
35. Cerqueira, M.B.R.; Caldas, S.S.; Primel, E.G. New sorbent in the dispersive solid phase extraction step of quick, easy, cheap, effective, rugged, and safe for the extraction of organic contaminants in drinking water treatment sludge. *J. Chromatogr. A* **2014**, *1336*, 10–22. [CrossRef] [PubMed]
36. Bhatnagar, A.; Sillanpää, M. Applications of chitin- and chitosan-derivatives for the detoxification of water and wastewater—A short review. *Adv. Colloid Interface Sci.* **2009**, *152*, 26–38. [CrossRef] [PubMed]

37. Cho, C.-W.; Lim, C.-R.; Cho, B.-G.; Mun, S.-B.; Choi, J.-W.; Zhao, Y.; Kim, S.; Yun, Y.-S. Development of prediction models for adsorption properties of chitin and chitosan for micropollutants. *Chem. Eng. J.* **2021**, *426*, 131341. [CrossRef]
38. Ravi Kumar, M.N.V. A review of chitin and chitosan applications. *React. Funct. Polym.* **2000**, *46*, 1–27. [CrossRef]
39. Si, R.; Han, Y.; Wu, D.; Qiao, F.; Bai, L.; Wang, Z.; Yan, H. Ionic liquid-organic-functionalized ordered mesoporous silica-integrated dispersive solid-phase extraction for determination of plant growth regulators in fresh panax ginseng. *Talanta* **2020**, *207*, 120247. [CrossRef]
40. Ścigalski, P.; Kosobucki, P. Recent materials developed for dispersive solid phase extraction. *Molecules* **2020**, *25*, 4869. [CrossRef]
41. Islam, A.K.M.M.; Hong, S.-M.; Lee, H.-S.; Moon, B.-C.; Kim, D.; Kwon, H. Identification and characterization of matrix components in spinach during QuEChERS sample preparation for pesticide residue analysis by LC–ESI–MS/MS, GC–MS and UPLC-DAD. *J. Food Sci. Technol.* **2018**, *55*, 3930–3938. [CrossRef] [PubMed]
42. Holmes, B.; Dunkin, A.; Schoen, R.; Wiseman, C. Single-laboratory ruggedness testing and validation of a modified QuEChERS-approach to quantify 185 pesticide residues in salmon by liquid chromatography- and gas chromatography-tandem mass spectrometry. *J. Agric. Food Chem.* **2015**, *63*, 5100–5106. [CrossRef]
43. Duedahl-Olesen, L.; Iversen, N.M.; Kelmo, C.; Jensen, L.K. Validation of QuEChERS for screening of 4 marker polycyclic aromatic hydrocarbons in fish and malt. *Food Control* **2020**, *108*, 106434. [CrossRef]
44. Pano-Farias, N.S.; Ceballos-Magaña, S.G.; Jurado, J.M.; Aguayo-Villarreal, I.A.; Muñiz-Valencia, R. Analytical method for pesticides in avocado and papaya by means of ultra-high performance liquid chromatography coupled to a triple quadrupole mass detector: Validation and uncertainty assessment. *J. Food Sci.* **2018**, *83*, 2265–2272. [CrossRef]
45. Vijaykumar, L.K.; Krishnegowda, D.N.; Judith, S.; Shankarappa, L.T.; Mayanna, A.; Sanganal, J.S.; Hegde, R. Method validation and quantification of 188 pesticides in animal liver using QuEChERS and GC-MS/MS. *Toxicol. Anal. Clin.* **2024**, *in press*. [CrossRef]
46. Gilbert-López, B.; García-Reyes, J.F.; Molina-Díaz, A. Sample treatment and determination of pesticide residues in fatty vegetable matrices: A review. *Talanta* **2009**, *79*, 109–128. [CrossRef] [PubMed]
47. Theurillat, X.; Dubois, M.; Huertas-Pérez, J.F. A multi-residue pesticide determination in fatty food commodities by modified QuEChERS approach and gas chromatography-tandem mass spectrometry. *Food Chem.* **2021**, *353*, 129039. [CrossRef]
48. Besil, N.; Cesio, V.; Heinzen, H.; Fernandez-Alba, A.R. Matrix effects and interferences of different citrus fruit coextractives in pesticide residue analysis using ultrahigh-performance liquid chromatography–high-resolution mass spectrometry. *J. Agric. Food Chem.* **2017**, *65*, 4819–4829. [CrossRef]
49. Rizzetti, T.M.; Kemmerich, M.; Martins, M.L.; Prestes, O.D.; Adaime, M.B.; Zanella, R. Optimization of a QuEChERS based method by means of central composite design for pesticide multiresidue determination in orange juice by UHPLC–MS/MS. *Food Chem.* **2016**, *196*, 25–33. [CrossRef]
50. Alcântara, D.B.; Fernandes, T.S.M.; Nascimento, H.O.; Lopes, A.F.; Menezes, M.G.G.; Lima, A.C.A.; Carvalho, T.V.; Grinberg, P.; Milhome, M.A.L.; Oliveira, A.H.B.; et al. Diagnostic detection systems and QuEChERS methods for multiclass pesticide analyses in different types of fruits: An overview from the last decade. *Food Chem.* **2019**, *298*, 124958. [CrossRef]
51. Kaczyński, P.; Łozowicka, B.; Perkowski, M.; Szabuńko, J. Multiclass pesticide residue analysis in fish muscle and liver on one-step extraction-cleanup strategy coupled with liquid chromatography tandem mass spectrometry. *Ecotox. Environ. Safe.* **2017**, *138*, 179–189. [CrossRef]
52. Cvetkovic, J.S.; Mitic, V.D.; Stankov Jovanovic, V.P.; Dimitrijevic, M.V.; Petrovic, G.M.; Nikolic-Mandic, S.D.; Stojanovic, G.S. Optimization of the QuEChERS extraction procedure for the determination of polycyclic aromatic hydrocarbons in soil by gas chromatography-mass spectrometry. *Anal. Methods* **2016**, *8*, 1711–1720. [CrossRef]
53. Sunyer-Caldú, A.; Diaz-Cruz, M.S. Development of a QuEChERS-based method for the analysis of pharmaceuticals and personal care products in lettuces grown in field-scale agricultural plots irrigated with reclaimed water. *Talanta* **2021**, *230*, 122302. [CrossRef]
54. Núñez, M.; Borrull, F.; Fontanals, N.; Pocurull, E. Determination of pharmaceuticals in bivalves using QuEChERS extraction and liquid chromatography-tandem mass spectrometry. *Anal. Bioanal. Chem.* **2015**, *407*, 3841–3849. [CrossRef]
55. Alharbi, O.M.L.; Basheer, A.A.; Khattab, R.A.; Ali, I. Health and environmental effects of persistent organic pollutants. *J. Mol. Liq.* **2018**, *263*, 442–453. [CrossRef]
56. Turusov, V.; Rakitsky, V.; Tomatis, L. Dichlorodiphenyltrichloroethane (ddt): Ubiquity, persistence, and risks. *Environ. Health Perspect.* **2002**, *110*, 125–128. [CrossRef] [PubMed]
57. Srogi, K. Monitoring of environmental exposure to polycyclic aromatic hydrocarbons: A review. *Environ. Chem. Lett.* **2007**, *5*, 169–195. [CrossRef] [PubMed]
58. MSAC—Germany. 1-[4-(1,1-dimethylethyl)phenyl]-3-(4-methoxyphenyl)propane-1,3-dione: Justification for the Selection of a Substance for CoRAP Inclusion. Available online: https://echa.europa.eu/documents/10162/ee03f3cf-f7b8-88ed-8d49-b613a921cf65 (accessed on 23 September 2024).
59. Campos-Mañas, M.C.; Ferrer, I.; Thurman, E.M.; Agüera, A. Opioid occurrence in environmental water samples—A review. *Trends Environ. Anal. Chem.* **2018**, *20*, e00059. [CrossRef]
60. Prosser, R.S.; Sibley, P.K. Human health risk assessment of pharmaceuticals and personal care products in plant tissue due to biosolids and manure amendments, and wastewater irrigation. *Environ. Int.* **2015**, *75*, 223–233. [CrossRef]
61. Bergmann, A.; Fohrmann, R.; Weber, F.-A. *Zusammenstellung von Monitoringdaten zu Umweltkonzentrationen von Arzneimitteln*; Umweltbundesamt: Dessau-Roßlau, Germany, 2011.
62. Downs, C.A.; Kramarsky-Winter, E.; Segal, R.; Fauth, J.; Knutson, S.; Bronstein, O.; Ciner, F.R.; Jeger, R.; Lichtenfeld, Y.; Woodley, C.M.; et al. Toxicopathological effects of the sunscreen UV filter, oxybenzone (benzophenone-3), on coral planulae and cultured

primary cells and its environmental contamination in Hawaii and the U.S. Virgin islands. *Arch. Environ. Contam. Toxicol.* **2016**, *70*, 265–288. [CrossRef] [PubMed]
63. U.S. Department of Agriculture Agricultural Research Service. Spinach, Mature. Available online: https://fdc.nal.usda.gov/fdc-app.html#/food-details/1999633/nutrients (accessed on 1 August 2024).
64. U.S. Department of Agriculture Agricultural Research Service. Avocado, Raw. Available online: https://fdc.nal.usda.gov/fdc-app.html#/food-details/1102652/nutrients (accessed on 1 August 2024).
65. U.S. Department of Agriculture Agricultural Research Service. Oranges, Raw, Navels. Available online: https://fdc.nal.usda.gov/fdc-app.html#/food-details/746771/nutrients (accessed on 1 August 2024).
66. U.S. Department of Agriculture Agricultural Research Service. Fish, Salmon, Atlantic, Farm Raised, Raw. Available online: https://fdc.nal.usda.gov/fdc-app.html#/food-details/2684441/nutrients (accessed on 1 August 2024).
67. U.S. Department of Agriculture Agricultural Research Service. Pork, Fresh, Variety Meats and By-Products, Liver, Raw. Available online: https://fdc.nal.usda.gov/fdc-app.html#/food-details/167862/nutrients (accessed on 1 August 2024).
68. European Commission. Analytical Quality Control and Method Validation Procedure for Pesticide Residues Analysis in Food and Feed SANTE/11312/2021. Available online: https://food.ec.europa.eu/system/files/2023-11/pesticides_mrl_guidelines_wrkdoc_2021-11312.pdf (accessed on 20 August 2024).
69. Codex Alimentarius, International Food Standards. Guidelines on Perfomance Criteria for Methods of Analysis for the Determination of Pesticide Residues in Food and Feed CXG 90-2017. Available online: https://www.fao.org/fao-who-codexalimentarius/sh-proxy/en/?lnk=1&url=https%253A%252F%252Fworkspace.fao.org%252Fsites%252Fcodex%252FStandards%252FCXG+90-2017%252FCXG_090e.pdf (accessed on 9 August 2024).
70. EU Reference Laboratory for Pesticides Requiring Single Residue Methods. Analysis of Ethylene Oxide and Its Metabolite 2-Chloroethanol by the Quoil or the QuEChERS Method and GC-MS/MS. Available online: https://www.eurl-pesticides.eu/library/docs/srm/EurlSrm_Observation_EO_V1.pdf (accessed on 12 August 2024).
71. Wang, J.; Zhang, J.; Han, L.; Wang, J.; Zhu, L.; Zeng, H. Graphene-based materials for adsorptive removal of pollutants from water and underlying interaction mechanism. *Adv. Colloid Interface Sci.* **2021**, *289*, 102360. [CrossRef]
72. Peter, M.; Bakanov, N.; Mathgen, X.; Brühl, C.A.; Veith, M.; Müller, C. Multiresidue analysis of bat guano using GC-MS/MS. *Anal. Bioanal. Chem.* **2024**, *416*, 3149–3160. [CrossRef]
73. EU Reference Laboratories for Residues of Pesticides—Single Residue Methods. Use of Analyte Protectants in GC-Analysis: A Way to Improve Peak Shape and Reduce Decomposition of Susceptible Compounds. Available online: https://www.eurl-pesticides.eu/library/docs/srm/EURL_Observation-APs.pdf (accessed on 13 August 2024).

molecules

MDPI

Article

Emulsive Liquid–Liquid Microextraction for the Determination of Phthalic Acid Esters in Environmental Water Samples

Xinyuan Bi [1,2,3], Chi Zhang [1,2], Xiaorong Xue [1,2], Shangjun Su [1,2], Zhiping Yang [1,2], Xu Jing [4] and Qiang Zhang [1,2,*]

1 College of Resources and Environment, Shanxi Agricultural University, Taigu 030801, China; bixinyuan98@163.com (X.B.); z591374178@163.com (C.Z.); r1044381617@163.com (X.X.); sxsushj@126.com (S.S.); yzpsx0208@163.com (Z.Y.)

2 Institute of Eco-Environment and Industrial Technology, Shanxi Agricultural University, Taiyuan 030031, China

3 College of Agricultural Economics and Management, Shanxi Agricultural University, Taiyuan 030006, China

4 College of Food Science and Engineering, Shanxi Agricultural University, Taigu 030801, China; x.jing@vip.163.com

* Correspondence: zhangqiang0351@163.com; Tel.: +86-0351-7639180

Abstract: A convenient, rapid, and environmentally friendly method, emulsive liquid–liquid microextraction combined with high-performance liquid chromatography, was established to determine phthalic acid esters in tap, river, lake, and sea water. After the method's optimization, we obtained the appropriate volume of the extractant and pure water, the number of strokes, the separation methods, the mass volume fraction of the demulsifier, the demulsifier volume, the sample volume, the salt amount, and the pH conditions. This method requires only 200 μL of heptanoic acid (fatty acid) as the extractant and 75 mg of sodium acetate as demulsifiers for fast microextraction and separation, respectively, avoiding the use of further equipment. Emulsive liquid–liquid microextraction offers substantial advantages over dispersive liquid–liquid microextraction by eliminating the need for toxic dispersants, thereby preventing any influences of dispersants on the partition coefficients. The linear range of detection ranged from 0.5 to 50 μg L^{-1}, with a limit of detection of 0.2 μg L^{-1} and a limit of quantitation of 0.5 μg L^{-1}. The recoveries ranged from 80.2% to 106.3%, and the relative standard deviations ranged between 0.5% and 6.7%. Five greenness metrics confirmed that this method is environmentally friendly and aligns with the principles of green analytical chemistry. The proposed method achieved a greenness score of 8.42, surpassing that of other methods as evaluated using the SPMS. The novel method may well be a valuable technique for determining phthalic acid esters in water samples.

Keywords: emulsive liquid–liquid microextraction; fatty acid; demulsifier; phthalic acid esters; greenness evaluation

1. Introduction

Phthalic acid esters (PAEs) consist of a benzene ring with two ester groups positioned in the ortho-configuration [1], which are used as plasticizers to improve the plasticity, durability, transparency, and strength of materials [2]. Every year, three million metric tons of phthalate derivatives are applied in various industrially manufactured products [3], such as food packaging, fertilizers, pesticides, detergents, and other industrial products. Dimethyl phthalate (DMP), diethyl phthalate (DEP), dipropyl phthalate (DPrP), dibutyl phthalate (DBP), and dipentyl phthalate (DPeP) are commonly used PAE compounds. Furthermore, the integration of PAEs into plastic products primarily relies on weak physical interactions, such as hydrogen bonding and van der Waals forces, rather than chemical bonds [4], which significantly contributes to their environmental release. Consequently, the widespread usage of PAEs has resulted in their effortless permeation into the environment through leaching, evaporation, and abrasion [5], resulting in the contamination of ecosystems. Studies show that PAEs harm human health [6], reducing sperm quality, causing infertility, disrupting the nervous system [7], and potentially increasing the risk of cancers, especially breast and liver cancer [8]. Indeed, PAEs could be widely detected in groundwater, river water, and seawater [9]; therefore, it is necessary to develop techniques for detecting PAEs in water samples.

As PAEs have inherently low solubility, their concentrations in environmental water samples are often trace-level, requiring sample pretreatment before chromatographic analysis. Liquid–liquid extraction (LLE) and solid-phase extraction (SPE) have been traditionally used as prevalent pretreatment techniques [10]. However, the methods are often time-consuming and require substantial amounts of organic solvents, posing limitations in terms of efficiency and environmental impact. To mitigate these inefficiencies, liquid-phase microextraction (LPME) and solid-phase microextraction (SPME) have emerged as innovative alternatives. LPME, a miniaturized version of LLE, utilizes mere microliters of solvent and hastens extraction speeds [11]. SPME minimizes the use of solid adsorbent but causes lags in LPME in terms of the extraction velocity. LPME boasts advantages such as simplicity of operation, shortened processing time, and cost-effectiveness. Nevertheless, its primary drawback lies in the utilization of toxic dispersants to facilitate the dispersion of the extractant within samples. Commonly used dispersants, including methanol, acetonitrile, and acetone, contribute to an increase in the overall toxic solvent load during the extraction process. Furthermore, the dissolution of these dispersants in samples diminishes the partition coefficient of analytes, thus impeding their efficient transfer to extractants.

Emulsive liquid–liquid microextraction (ELLME) was reported (in 2024) as a novel extraction technique that does not use toxic dispersants [12]. Its fundamental principle involves the mixing of an extractant with water to create a concentrated primary oil-in-water (O/W) emulsion; it is then rapidly injected into the sample, resulting in the formation of a diluted secondary O/W emulsion. This diluted emulsion, composed of the extractant and sample, facilitates full contact due to its large interfacial area, enabling the rapid and efficient extraction of analytes in samples [13]. Compared to previously reported LPME, ELLME based on O/W emulsions offers several advantages, including shorter operation time, milder extraction conditions, lower analyte losses, and the elimination of the need for additional toxic reagents or specialized equipment [14].

In LPME, the choice of extractant is paramount, as it directly affects both the extraction efficiency and the recovery of target analytes [15]. Fatty acids derived from natural resources are non-toxic or of low toxicity to organisms, have good biocompatibility, and can efficiently separate target analytes, aligning with the principles of green chemistry and sustainable development [16]. Fatty acids may be used as extractants, which avoids the use of traditional organic solvents and reduces environmental pollution. Heptanoic acid stands out as a promising choice, offering advantages, such as high extraction efficiency, environmental protection potential, selective extraction ability, and fast extraction kinetics [17]. Heptanoic acid can ionize to form hydrophilic anions and possesses the properties of an anionic surfactant, enabling it to extract hydrophobic analytes into its formed micelles, thereby significantly improving extraction efficiency [18]. This innovative emulsive microextraction mechanism based on heptanoic acid can accelerate the rate of analyte allocation between phases, ultimately facilitating efficient rapid extraction. The type and volume of the extractant [19], the separation technology utilized, the sample volume, the salt amount, and the pH can impact the extraction efficiency and enrichment factors [15].

Phase separation not only affects the extraction efficiency of analytes but also influences the simplicity of extraction operation and the degree of environmental pollution. Therefore, continuous development of phase separation methods is necessary in the application and development of LPME. Centrifugation, as the convention phase separation method, is often time-consuming and resource-intensive, prolonging the process and requiring extra equipment. Demulsifiers characterized by high surface activity and low surface tension are potentially powerful separation agents, enabling rapid demulsification and phase separation [20]. Sodium acetate is a biodegradable and cost-effective chemical, aligning with the principles of green chemistry. The hydroxyl part of the acetate ion forms a hydrogen bond with water molecules, creating a hydrophilic head, while the sodium ion behaves as a cation that can potentially aid in the transfer of target analytes from samples to the extractant [21], thereby potentially increasing the extraction efficiency of LPME processes.

The present work aims to develop a novel method to determine five PAEs (i.e., DMP, DEP, DPrP, DBP, and DPeP) in water samples by ELLME using fatty acids. Also, some variables (e.g., the extractant volume, pure water volume, number of strokes, separation methods, mass volume fraction of the demulsifier, demulsifier volume, sample volume, salt amount, and pH) will be tested to improve the recovery of PAEs. Finally, the applicability of the method will be performed by analyzing real tap, river, lake, and sea water samples.

2. Results and Discussion

2.1. Optimization of the Extraction Procedure

Tap water was used as the water sample in the subsequent optimization test, and a concentration of 5 μg L^{-1} was spiked into the blank matrix of tap water (Figure S1). The recovery of PAEs was affected by an array of ELLME factors, including the extractant volume, the pure water volume, the number of strokes, the separation methods, the mass volume fraction of the demulsifier, the demulsifier volume, the sample volume, the amount of sodium acetate, and the pH. To refine and maximize the recovery, optimization experiments were designed to optimize one variable at a time, performing each extraction condition in triplicates. The initial parameters were established as follows: 200 μL of extractant and 600 μL of pure water were used, and mixed 8 times. The sample volume was set to 7 mL, with 500 μL of a 15% (w/v) sodium acetate trihydrate solution added as the demulsifier. No additional salt was added, and the pH was left unadjusted.

The method for calculating the extraction recovery is outlined below:

$$\text{Extraction recovery } (\%) = \frac{C_{\text{measured}} - C_{\text{real}}}{C_{\text{spiked}}} \times 100\%$$

$C_{measured}$, C_{real}, and C_{spiked} represent the measured concentration of target analytes in spiked samples, the concentration of target analytes in real samples, and the spiked concentration of target analytes, respectively.

2.1.1. Optimization of the Extractant Volume

The proper extractant volume is conducive to the separation of the extractant from the sample and ensures method sensitivity [19]. The effect of the extractant volume on the recovery range was studied with volumes ranging from 140 to 220 μL (Figure 1A). When the extractant volume increased from 140 to 200 μL, the recovery increased accordingly. This may be because an insufficient extractant cannot fully extract target analytes. When 200 μL was used, the recovery reached its peak and gradually decreased with further increases in the volume. It may be that excessive extractant affected the emulsion's formation. Therefore, 200 μL of the extractant was chosen for further optimization in subsequent tests.

2.1.2. Optimization of the Pure Water Volume

ELLME involves adding a concentrated O/W emulsion, which consists of heptanoic acid and pure water, to the sample, forming an emulsion system, and completing the extraction. The volume of pure water affects the emulsion concentration and recovery efficiency. The influence of the pure water volume on the recovery range was assessed with volumes ranging from 0 to 500 μL (Figure 1B). The best effect was achieved at 400 μL. An emulsion formulated with an optimal volume ratio may potentially enlarge the contact area between the extractant and analytes [13], thereby improving the recovery. When the volume of water was low, it was difficult for the emulsion system to form a stable O/W emulsion, and the inversion of the emulsion was more likely to occur. The high pure water volume led to low-concentration emulsion droplets. Therefore, 400 μL of the pure water volume was selected for further optimization in subsequent tests.

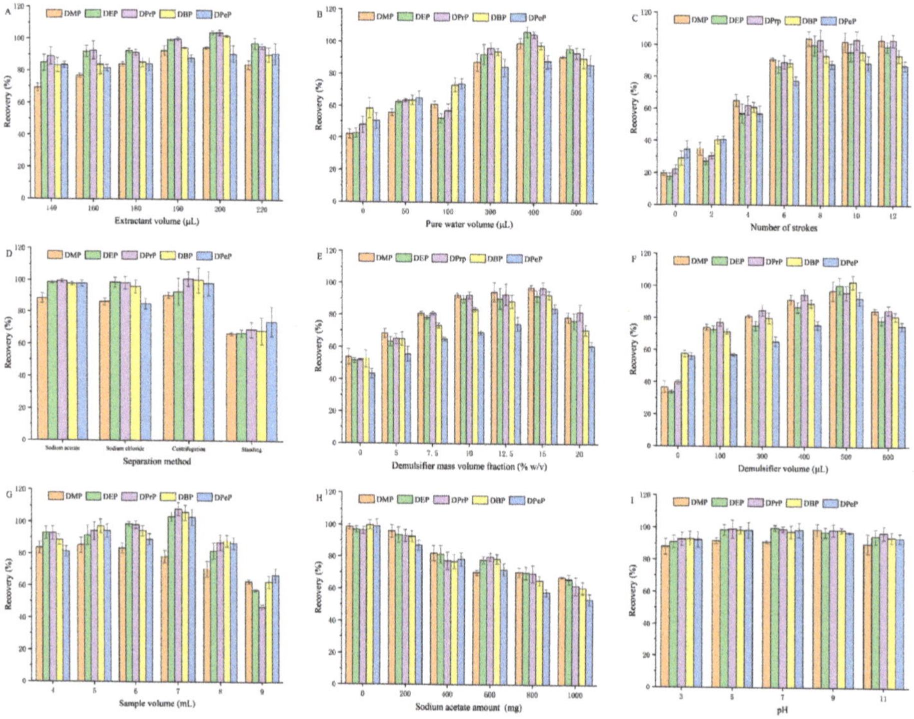

Figure 1. Optimization of ELLME parameters: (**A**) extractant volume; (**B**) pure water volume; (**C**) number of strokes; (**D**) separation method; (**E**) demulsifier mass volume fraction (% w/v); (**F**) demulsifier volume; (**G**) sample volume; (**H**) sodium acetate amount; (**I**) pH (the error bar represents the standard deviation).

2.1.3. Optimization of the Number of Strokes

The number of strokes affects the formation of the concentrated O/W emulsion and extraction efficiency. The effect of the number of strokes on the recovery range was studied with numbers ranging from 0 to 12 times (Figure 1C). The recovery increased with the increase in the number of strokes and remained unchanged after reaching eight strokes. It is likely that a certain number of strokes are required to sufficiently mix heptanoic acid and pure water, thereby ensuring the formation of the emulsion. Thus, eight cycles were selected for further optimization in subsequent tests.

2.1.4. Optimization of Separation Methods

The choice of separation method can directly affect the extraction efficiency and time. Our study adopted the following separation techniques: adding organic salt (15% w/v sodium acetate) or inorganic salt (15% w/v sodium chloride) serving as demulsifiers, centrifugation (4000 rpm for 5 min), and standing for 30 min (Figure 1D). The recovery in sodium acetate and centrifugation treatment exceeds that of sodium chloride and standing treatment. The interfacial film at the O/W interface is more susceptible to disruption by external stimuli or droplet collisions, facilitating droplet agglomeration and enhancing the separation process [22]. Sodium acetate demonstrates remarkable surface activity and demulsification capacity, which can be used to exclude the use of a centrifuge and reduce the separation time. Its primary role is to diminish the free water surrounding extractant droplets via salt precipitation, thereby destabilizing the emulsion. Therefore, sodium acetate was selected as the demulsifier for further optimization.

2.1.5. Optimization of the Mass Volume Fraction of Demulsifier

The mass volume fraction of the demulsifier determines its demulsification performance. The influence of the mass volume fraction of the demulsifier on the recovery range was estimated, ranging from 0 to 20% (Figure 1E). Within a certain range (0–15%), the stability of the emulsion system is generally enhanced with the increase in the mass volume fraction of the demulsifier. A moderate amount of sodium acetate can promote the breaking of the emulsion by altering the properties of the emulsion interface, such as reducing interfacial tension or increasing collision frequency. However, when the mass volume fraction of the demulsifier exceeds 15%, an excessive amount of demulsifier may increase the viscosity of the sample and alter the chemical environment of the emulsion, hindering interactions between the droplets. This is detrimental to breaking and subsequently to the emulsion, leading to a lower extraction efficiency. Therefore, 15% of the mass volume fraction was selected as the appropriate mass volume fraction of the demulsifier.

2.1.6. Optimization of the Demulsifier Volume

As the volume of the demulsifier affects the phase separation time and efficiency, the influence of the demulsifier volume on the recovery range was studied across a range from 0 to 900 μL (Figure 1F). A large number of demulsifiers may be able to make fuller contact with the emulsion [23], thereby affecting the recovery. However, as the demulsifier volume exceeds a certain range, the interaction between the molecules of the demulsifier is enhanced, leading to heightened dissolution and consequently diminished demulsification effectiveness. The recovery was found to peak at 500 μL; thus, 500 μL of demulsifier was chosen as the optimal demulsifier volume.

2.1.7. Optimization of the Sample Volume

The distribution coefficient of target analytes in the sample affects the mass transfer and then affects its recovery. The influence of the sample volume on the recovery range was investigated across a range from 4 to 9 mL (Figure 1G). The highest recovery was observed at 7 mL. Therefore, a sample volume of 7 mL was selected for the subsequent experiments.

2.1.8. Optimization of the Salt Amount

To reduce the solubility in samples and improve extraction efficiency, adding a certain amount of salt to the sample may alter the partition between organic and water phases. The influence of the initial sodium acetate amount on the recovery was studied across a range from 0 to 1000 mg (Figure 1H). With the increase in the amount of initial salt added, the recovery decreased; this might be because excessive sodium acetate affected the partition coefficient between organic and water phases, thereby influencing the extraction efficiency and the recovery. Excessive amounts of sodium acetate may also increase the viscosity and density of the sample, reducing the extraction efficiency [24]. Therefore, additional sodium acetate was not added in subsequent experiments.

2.1.9. Optimization of the pH of the Sample

The optimal pH value of the sample affects the analyte conversion into their neutral forms and the extraction efficiency [25]. The initial pH of the sample was 7.94; to investigate the effect of sample pH on the extraction efficiency, the pH values of samples were adjusted by 1 mol L^{-1} HCl and NaOH solution. pH values of 3, 5, 7, 9, and 11 were studied (Figure 1I). The extraction efficiency remained high at each pH value tested; thus, the experiment proceeded without any modification to the pH of the sample.

2.2. Validation Parameters of the Proposed Methodology

Tap, river, lake, and sea water were extracted using the optimal ELLME-HPLC method developed in this study. Mixed standard working solutions of various concentrations were prepared and spiked into the blank matrix of tap, river, lake, and sea water (Figure S1). The concentrations of the spiked standards were 0.5, 1, 5, 10, 50, and 100 μg L^{-1}. Quantitation was performed using the external standard method. For each substrate (tap, river, lake, and sea water), the concentration of each PAE in the spiked samples was denoted by x, and the corresponding peak area obtained through detection was denoted by y. The linear equation, coefficient of determination (R^2), limit of detection (LOD), limit of quantitation (LOQ), and intra-day and inter-day relative standard deviations (RSDs) for PAEs in each substrate were evaluated. The R^2 values for all PAEs were greater than 0.998 (Table 1). The calculation of LODs and LOQs was based on a signal-to-noise ratio of 3 to 10 times. The LOD and LOQ were found to be 0.2 and 0.5 μg L^{-1}, respectively. The intra-day (five times on the same day) and inter-day RSDs (five times on consecutive days) for each PAE ranged from 0.3% to 6.1% and from 1.1% to 9.1%, respectively.

Table 1. Assessment parameters of the proposed technique for detecting PAEs.

PAE	Sample	Linear Equation (μg L^{-1})	R^2	LOD (μg L^{-1})	LOQ (μg L^{-1})	Intra-Day RSD (%)	Inter-Day RSD (%)	Extraction Recovery		
								0.5 μg L^{-1}	5 μg L^{-1}	50 μg L^{-1}
DMP	Tap	$y = 25187.02x + 4026.76$	0.999	0.2	0.5	4.4	4.7	91.8 ± 3.5	80.2 ± 2.5	83.1 ± 2.7
	River	$y = 25697.58x + 7882.46$	0.998	0.2	0.5	3.9	7.4	96.4 ± 4.2	87.2 ± 1.8	86.2 ± 1.3
	Lake	$y = 25748.44x + 4358.51$	0.999	0.2	0.5	2.8	1.8	88.0 ± 3.3	84.2 ± 2.1	83.2 ± 0.6
	Sea	$y = 24738.13x + 4654.33$	0.999	0.2	0.5	2.7	8.6	81.3 ± 2.1	81.5 ± 2.9	83.3 ± 1.6
DEP	Tap	$y = 20798.26x - 2624.04$	0.999	0.2	0.5	6.1	6.5	97.9 ± 2.6	89.9 ± 1.6	96.5 ± 2.2
	River	$y = 21377.78x + 6516.51$	0.998	0.2	0.5	1.5	6.9	99.4 ± 4.3	95.3 ± 5.4	99.3 ± 1.7
	Lake	$y = 21329.38x + 6771.08$	0.999	0.2	0.5	3.6	2.1	94.8 ± 4.1	95.3 ± 1.4	98.6 ± 3.1
	Sea	$y = 20415.78x + 5976.21$	0.999	0.2	0.5	2.3	2.4	103.9 ± 3.0	84.7 ± 5.3	99.9 ± 1.7
DPrP	Tap	$y = 17303.09x - 6085.57$	0.999	0.2	0.5	5.0	9.1	102.8 ± 1.6	91.0 ± 0.6	98.3 ± 1.7
	River	$y = 17882.65x - 555.45$	0.999	0.2	0.5	2.1	6.0	97.6 ± 4.6	95.6 ± 3.6	102.3 ± 0.8
	Lake	$y = 17758.35x + 2071.82$	0.999	0.2	0.5	1.4	1.1	92.8 ± 4.5	97.1 ± 2.2	100.0 ± 3.2
	Sea	$y = 17029.26x - 307.17$	0.999	0.2	0.5	3.4	6.3	98.4 ± 3.3	88.5 ± 2.1	101.3 ± 1.9
DBP	Tap	$y = 19974.01x - 2334.41$	0.999	0.2	0.5	2.6	1.4	101.4 ± 5.0	96.7 ± 0.5	98.6 ± 1.6
	River	$y = 20602.78x + 4967.80$	0.998	0.2	0.5	1.2	5.1	96.1 ± 6.4	101.2 ± 1.9	102.5 ± 0.7
	Lake	$y = 20508.70x + 6984.56$	0.998	0.2	0.5	0.3	1.1	95.7 ± 3.9	99.7 ± 4.8	100.4 ± 3.2
	Sea	$y = 19648.01x + 4635.30$	0.999	0.2	0.5	4.8	8.8	97.1 ± 0.8	104.6 ± 3.1	101.9 ± 2.0
DPeP	Tap	$y = 18080.12x - 4170.32$	0.999	0.2	0.5	1.9	7.2	100.3 ± 6.1	96.6 ± 0.9	97.9 ± 1.6
	River	$y = 18603.57x + 3207.69$	0.999	0.2	0.5	1.1	5.4	80.9 ± 5.3	101.3 ± 1.2	101.2 ± 0.8
	Lake	$y = 18494.86x + 4973.91$	0.998	0.2	0.5	0.8	2.6	85.8 ± 3.5	101.6 ± 6.0	99.2 ± 3.2
	Sea	$y = 17767.28x + 2667.39$	0.999	0.2	0.5	4.0	9.0	80.8 ± 1.3	106.3 ± 2.7	100.6 ± 1.9

To assess the suitability of the ELLME-HPLC method for detecting PAEs, initial tests on real tap, river, lake, and sea water samples were conducted. The results showed that these samples were free of detectable levels of PAEs. Subsequently, to further validate the method, spiked samples of tap, river, lake, and sea water with known concentrations of PAEs (0.5, 5, and 50 μg L^{-1}, respectively) were spiked for analysis. As shown in Table 1, the recoveries obtained through standard addition ranged from 80.2% to 106.3%, and RSDs ranged from 0.5% to 6.7%. The results indicate that ELLME-HPLC is reliable and can be used to detect PAE residues in various water samples.

2.3. Evaluation of Eco-Profile

The Analytical Eco-scale (AES), the Green Analytical Procedure Index (GAPI), Analytical GREEnness (AGREE), the Analytical GREEnness metric for sample preparation (AGREEprep), and the sample preparation metric of sustainability (SPMS) were all considered when designing the methodology for evaluating greenness in this study.

AES as a semi-quantitative tool evaluates the environmental friendliness of an analytical method over its entire life cycle by considering factors such as the hazards posed by the reagents, the amount of reagents consumed, the energy consumption, and the amount of waste produced [26]. The evaluation is based on a maximum score of 100, with penalty points deducted for each criterion that does not meet ideal environmental standards. A higher score indicates a more environmentally friendly method. In general, a score of 75 or above is considered excellent green. In this instance, the method scored 85 (Tables 2 and S1), suggesting that it is green.

Table 2. The results of five green evaluation methodologies.

AES		GAPI	AGREE	AGREEprep	SPMS
	PPs				
1. Reagents	6	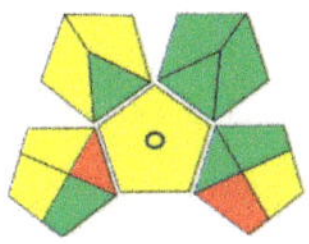		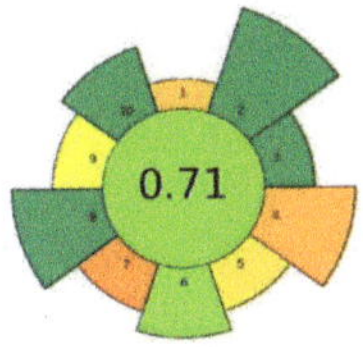	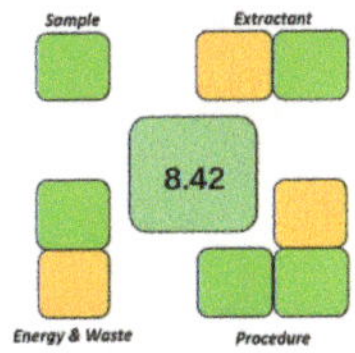
1.1 Acetonitrile	4				
1.2 Heptanoic acid	1				
1.3 Sodium acetate	1				
2. Instruments-HPLC	0				
3. Waste	9				
Total PPs	15				
Eco-scale score	85				

GAPI also estimates the environmental friendliness of an analytical method by considering multiple aspects, including sample preparation, reagent usage, waste generation, and energy consumption [27]. This assessment tool uses pentagons, incorporating a color-coding system with three distinct hues, green, yellow, and red, to signify the level of environmental friendliness of the analytical technique. Upon evaluating the test method, it was found that seven aspects were rated green, six were rated yellow, and two were rated red (Table 2). Notably, both the amount of reagent used and the safety aspects were deemed green. Overall, the developed method was considered relatively environmentally friendly.

AGREE is a comprehensive evaluation method in green analytical chemistry that is specifically designed to quantify the greenness of sample preparation and analytical procedures [28]. It evaluates the environmental performance by considering green chemistry principles, including reagent dosage and toxicity, waste generation, energy requirements, the number of steps involved, miniaturization, and automation. Upon completion of the evaluation, the method received a final AGREE score of 0.68 (Table 2), indicating a relatively low negative environmental impact.

AGREEprep is an assessment tool grounded in green sample preparation (GSP) that is a methodology specifically designed to assess the greenness of sample preparation processes. Its primary objective is to assess the environmental impact of sample preparation methods. AGREEprep takes into account factors such as reagent dosage, waste generation, and energy efficiency during the sample preparation phase [29]. In this instance, 200 μL of heptanoic acid and 75 mg sodium acetate trihydrate were used in the process of sample preparation. Upon evaluation, the method received a final score of 0.71 (Tables 2 and S2), indicating its high environmental friendliness and compliance with the GSP standard.

SPMS allows for a comprehensive examination of various factors involved in the sample preparation process, including resource consumption, energy usage, and pollutant emissions, thereby enabling a more precise assessment of its sustainability [30]. This methodology can help quantitatively analyze the sustainability of the sample preparation process, rendering the results more objective, detailed, and conducive to comparison. With an evaluation score of 8.42 (Table 2), the method shows a high level of environmental friendliness.

2.4. Comparison with Other Techniques

As shown in Table 3, the method proposed was compared with other liquid phase extraction methods for the chromatography analysis of PAEs, focusing on the analyte, environmental sample, pretreatment method, extraction reagent, extraction and separation device, detection instrument, LOQ, LOD, and ER. In the present study, five PAEs in four environmental water samples were detected. Only 200 μL of heptanoic acid and 75 mg of sodium acetate were used as these were considered green resources. In the reported cases, toxic and harmful reagents were used (e.g., 48 μL of dibutylamine, 200 mg of cresol, 20 μL of butyl lactate, and 400 μL of acetonitrile). Notably, the ELLME-HPLC method does not require the use of an ultrasonic bath, water bath, agitator, vortex mixer, centrifuge, or other equipment, which are typically required in the reported methods (e.g., ultrasonic bath for 10 min, water bath at 70 °C, vortex mixer for 1.8 min, centrifuge for 8 min, and agitator for 20 min). The LOQ and LOD are lower than the reported methods. The sample preparation metric of sustainability was adopted as a green evaluation methodology to evaluate the reported methods. The scores were 6.74 [31], 7.26 [32], 5.28 [33], and 5.58 [34], which are all lower than that of the proposed method, which was 8.42. In summary, ELLME has certain advantages in terms of greenness, simplicity, and sensitivity compared with other liquid phase extraction methods of PAEs.

Table 3. Comparisons of ELLME with other liquid phase extraction methods for chromatography analysis of PAEs.

PAE	Environmental Sample	Pretreatment Method	Extraction Reagent	Extraction and Phase Separation Device	Detection Instrument	LOQ ($\mu g\ L^{-1}$)	LOD ($\mu g\ L^{-1}$)	EF	ER (%)	SPMS Score	Reference
Dibutyl phthalate Benzyl butyl phthalate	River water Lake water	LLE	Tetrabutylammonium iodide + Decanoic acid (3000 μL)	Ultrasonic bath (10 min) Water bath (70 °C)	HPLC-UV	1.0–1.6	0.3–0.9	-	90–98	6.74	[31]
Diethyl phthalate Dibutyl phthalate Dioctyl phthalate Diethylhexyl phthalate	Tap water	HLLME	Dibutylamine (48 μL) Acetic acid (10 μL) Sodium carbonate (250 mg)	Agitator (0.5 min)	GC-FID	11.8–41.3	3.6–12.5	196–394	39–78	7.26	[32]
Dimethyl phthalate Diethyl phthalate Dibutyl phthalate Dipentyl phthalate	Tap water	DLLME	Cresol (200 mg) Butyl lactate (20 μL)	Vortex mixer (1.8 min) Ultrasonic bath (8 min) Centrifuge (8 min)	GC-FID	1–4	0.5–2	256–467	81.8–113.1	5.68	[33]
Didecyl phthalate	Sea water	DLLME	Octanol (500 μL) Acetonitrile (400 μL)	Agitator (20 min) Centrifuge (5 min)	HPLC-DAD	59	20	25	84.5–93.6	5.58	[34]
Dimethyl phthalate Diethyl phthalate Dipropyl phthalate Dibutyl phthalate Dipentyl phthalate	Tap water River water Lake water Sea water	ELLME	Heptanoic acid (200 μL) Sodium acetate (75 mg)	/	HPLC-UV	0.5	0.2	35	80.2–106.3	8.42	This Study

HLLME: homogeneous liquid–liquid microextraction. DLLME: dispersive liquid–liquid microextraction. GC-FID: gas chromatography–flame ionization detection.

3. Materials and Methods

3.1. Reagents, Chemicals, and Sample Preparation

DMP, DEP, DPrP, DBP, DPeP, heptanoic acid (98.0%), sodium acetate trihydrate (99.5%), sodium chloride (99.0%), acetonitrile (99.9%), and sodium hydroxide (99%) were obtained from Macklin Inc (Shanghai, China). Hydrochloric acid (37%) was obtained from Aladdin Co., Ltd., (Shanghai, China). Pure water was sourced from Barnstead™ GenPure™ Pro (Atvidaberg, Sweden). A stock solution of PAEs was prepared in acetonitrile at a concentration of 0.001g mL^{-1} and stored at 4 °C.

River and lake water were sourced from the Fenhe River and Jinyang Lake (Taiyuan, China), respectively. Sea water was sourced from the East China Sea (Xiamen, China), and the salinity was 29.8 g L^{-1}. The samples were filtered through a 0.22 μm filter membrane.

3.2. Instrumental

Chromatographic separations were conducted using an ACQUITY Arc HPLC system with a Waters 2489 UV/Vis detector. A Waters XBridge C_{18} column (4.6 × 250 mm, 5 μm) was employed at a temperature of 30 °C. The mobile phase was a mixture of acetonitrile and pure water, with an initial composition of 60:40 (v/v) and a final composition of 80:20 (v/v) at 10 min and a flow rate of 0.8 mL min^{-1}. The retention times of DMP, DEP, DPrP, DBP, and DPeP were 4.1, 5.6, 8.9, 13.4, and 18.8 min (Figure S1), respectively, with a wavelength set at 245 nm for detection.

3.3. ELLME-HPLC Protocol

At first, 200 μL of heptanoic acid and 400 μL of pure water in a 1.5 mL centrifugal tube were mixed eight times in 8 s with a plastic pipette tip to form a concentrated O/W emulsion. Then, the prepared emulsion was added to a glass centrifugal tube with 7 mL of a water sample to prepare a diluted O/W emulsion to complete the extraction. Finally, 500 μL of 15% sodium acetate solution as a demulsifier was added to fulfill phase separation within 2 min, and heptanoic acid was collected for analysis (Figure 2).

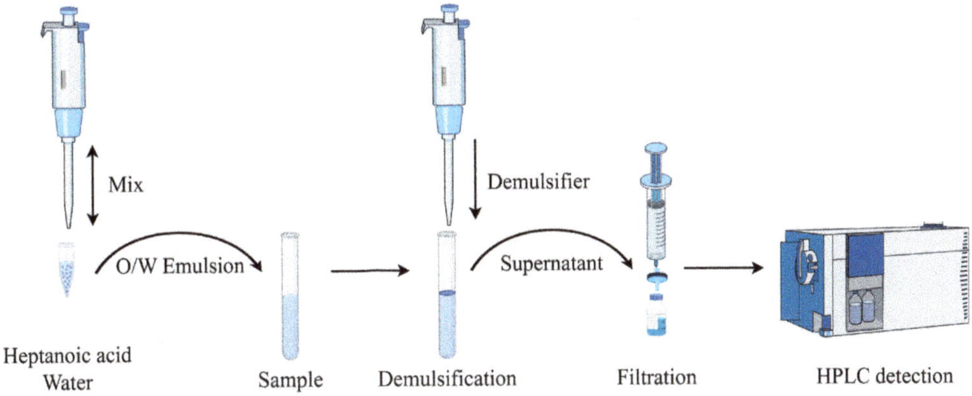

Figure 2. The flowchart of the ELLME-HPLC-UV/Vis by Figdraw 2.0 software.

4. Conclusions

In this study, a simple, rapid, and environmentally friendly, method was developed to determine the amount of PAEs in tap, river, lake, and sea water. The renewable and environmentally friendly heptanoic acid was used as an extractant. It was mixed with pure water to prepare a concentrated O/W emulsion and then added to the sample to form a diluted O/W emulsion to complete the fast extraction. The organic salt sodium acetate was then added as a demulsifier to complete rapid phase separation (instead of using a centrifuge). The main advantages include the avoidance of toxic extraction solvents, operational simplicity, cost-effectiveness, and the elimination of the need for additional auxiliary equipment. Evaluation using AES, GAPI, AGREE, AGREEprep, and SPMS showed that the method aligns with the development principles of green analytical chemistry and demonstrates environmental friendliness. The proposed method utilizes 200 μL of heptanoic acid and 400 μL of pure water, which are mixed with eight strokes. Subsequently, 500 μL of a 15% (w/v) sodium acetate solution is added as a demulsifier, enabling the processing of a 7 mL water sample without the need for additional salt or pH adjustment. In conclusion, this method holds significant potential as an effective tool for the sustainable green extraction and chromatographic determination of PAEs in environmental water samples.

Supplementary Materials: The following supporting information can be downloaded at: https://www.mdpi.com/article/10.3390/molecules29245908/s1. Figure S1: Chromatograms of the water samples. (A) PAE standards at a concentration of 5 μg L^{-1}; (B) tap water; (C) river water; (D) lake water; (E) sea water. 1. Dimethyl phthalate (DMP), 2. diethyl phthalate (DEP), 3. dipropyl phthalate (DPrP), 4. dibutyl phthalate (DBP), and 5. dipentyl phthalate (DPeP). Figure S2: The sample preparation metric of sustainability (SPMS) scores for reported methods. (A) literature [31]; (B) literature [32]; (C) literature [33]; (D) literature [34]. Table S1: Preparation of four environmental water samples. Table S2: Greenness assessment of the ELLME-HPLC technique using AES. Table S3: Greenness assessment of the ELLME-HPLC technique using AGREEprep.

Author Contributions: Conceptualization, X.J.; methodology, X.B.; software, X.X.; validation, S.S.; formal analysis, Z.Y.; investigation, X.B.; data, C.Z.; supervision, Q.Z. All authors have read and agreed to the published version of the manuscript.

Funding: This research was funded by [Distinguished and Excellent Young Scholar Cultivation Project of Shanxi Agricultural University] grant number [2023YQPYGC02].

Institutional Review Board Statement: Not applicable.

Informed Consent Statement: Not applicable.

Data Availability Statement: The data can be found in Supplementary Materials.

Acknowledgments: Not applicable.

Conflicts of Interest: The authors declare no conflicts of interest.

References

1. Sokołowski, A.; Kończak, M.; Oleszczuk, P.; Gao, Y.; Czech, B. Environmental and Food Contamination by Phthalic Acid Esters (PAEs): Overview. *Water Air Soil Pollut.* **2024**, *235*, 313. [CrossRef]
2. Huang, L.; Zhu, X.; Zhou, S.; Cheng, Z.; Shi, K.; Zhang, C.; Shao, H. Phthalic Acid Esters: Natural Sources and Biological Activities. *Toxins* **2021**, *13*, 495. [CrossRef] [PubMed]
3. Zhang, Y.; Jiao, Y.; Li, Z.; Tao, Y.; Yang, Y. Hazards of Phthalates (PAEs) Exposure: A Review of Aquatic Animal Toxicology Studies. *Sci. Total Environ.* **2021**, *771*, 145418. [CrossRef]
4. Wang, Z.; Ma, J.; Wang, T.; Qin, C.; Hu, X.; Mosa, A.; Ling, W. Environmental Health Risks Induced by Interaction between Phthalic Acid Esters (PAEs) and Biological Macromolecules: A Review. *Chemosphere* **2023**, *328*, 38578. [CrossRef]
5. Paluselli, A.; Kim, S.-K. Horizontal and Vertical Distribution of Phthalates Acid Ester (PAEs) in Seawater and Sediment of East China Sea and Korean South Sea: Traces of Plastic Debris? *Mar. Pollut. Bull.* **2020**, *151*, 110831. [CrossRef]
6. Wang, X.; Xu, M.; Yang, A.; Wang, Y.; Hou, S.; Zheng, N.; Liang, D.; Hua, X.; Dong, D. Health Risks of Population Exposure to Phthalic Acid Esters through the Use of Plastic Containers for Takeaway Food in China. *Sci. Total Environ.* **2021**, *785*, 147347. [CrossRef]
7. Xu, M.-L.; Cheng, Y.; Feng, M.; Lu, Q.; Lian, Y. Identifying Potential Sources of Phthalate Contamination in the Leaves of Stevia Rebaudiana (Bertoni) and the Development of Removal Technology. *Molecules* **2024**, *29*, 1627. [CrossRef]
8. Chang, W.-H.; Herianto, S.; Lee, C.-C.; Hung, H.; Chen, H.-L. The Effects of Phthalate Ester Exposure on Human Health: A Review. *Sci. Total Environ.* **2021**, *786*, 147371. [CrossRef]
9. Liang, J.; Ji, X.; Feng, X.; Su, P.; Xu, W.; Zhang, Q.; Ren, Z.; Li, Y.; Zhu, Q.; Qu, G.; et al. Phthalate Acid Esters: A Review of Aquatic Environmental Occurrence and Their Interactions with Plants. *J. Hazard. Mater.* **2024**, *470*, 134187. [CrossRef]
10. Vichapong, J.; Moyakao, K.; Kachangoon, R.; Burakham, R.; Santaladchaiyakit, Y.; Srijaranai, S. β-Cyclodextrin Assisted Liquid-Liquid Microextraction Based on Solidification of the Floating Organic Droplets Method for Determination of Neonicotinoid Residues. *Molecules* **2019**, *24*, 3954. [CrossRef]
11. Pena-Pereira, F.; Lavilla, I.; Bendicho, C. Liquid-Phase Microextraction Techniques within the Framework of Green Chemistry. *TrAC Trends Anal. Chem.* **2010**, *29*, 617–628. [CrossRef]
12. Guo, X.; Zheng, X.; Guo, X.; Wu, J.; Jing, X. Determination of Chiral Prothioconazole and Its Chiral Metabolite in Water, Juice, Tea, and Vinegar Using Emulsive Liquid-Liquid Microextraction Combined with Ultra-High Performance Liquid Chromatography. *Food Chem.* **2024**, *440*, 138314. [CrossRef] [PubMed]
13. Liu, J.; Pan, W.; Pei, T.; Wang, F.; Zhao, W.; Wang, E.; Li, L.; Jing, X. High-Throughput Semi-Automated Emulsive Liquid-Liquid Microextraction for Detecting SDHI Fungicides in Water, Juice, and Alcoholic Beverage Samples via UHPLC-MS/MS. *Talanta* **2024**, *274*, 126038. [CrossRef] [PubMed]
14. Liu, J.; Nie, Y.; Niu, Y.; Huang, L.; Jing, X. Lignin-Based Emulsive Liquid-Liquid Microextraction for Detecting Triazole Fungicides in Water, Juice, Vinegar, and Alcoholic Beverages via UHPLC-MS/MS. *Food Chem.* **2024**, *459*, 140407. [CrossRef]
15. Smith, C.D.; Downs, R.P.; Carrick, J.D.; Dietz, M.L. Determination of Extractant Solubility in Ionic Liquids by Thermogravimetric Analysis. *Solvent Extr. Ion Exch.* **2018**, *16*, 304–314. [CrossRef]
16. Gao, M.; Wang, J.; Zhang, X.; Dahlgren, R.A.; Rua, S.; Wang, X. Integrated Disperser Freezing Purification with Extraction Using Fatty Acid-Based Solidification of Floating Organic-Droplet (IDFP-EFA-SFO) for Triclosan and Methyltriclosan Determination in Seawater, Sediment and Seafood. *Mar. Pollut. Bull.* **2018**, *137*, 667–687. [CrossRef]
17. IJmker, H.M.; Gramblička, M.; Kersten, S.R.A.; Van Der Ham, A.G.J.; Schuur, B. Acetic Acid Extraction from Aqueous Solutions Using Fatty Acids. *Sep. Purif. Technol.* **2014**, *125*, 256–263. [CrossRef]
18. Glinski, J.; Chavepeyer, G.; Platten, J.-K. Adsorption of Heptanol, Heptanoic Acid, and Heptanoate Anions at Water-Air Interfaces. *J. Colloid Interface Sci.* **1996**, *180*, 290–292. [CrossRef]
19. Kannouma, R.E.; Hammad, M.A.; Kamal, A.H.; Mansour, F.R. Miniaturization of Liquid-Liquid Extraction; the Barriers and the Enablers. *Microchem. J.* **2022**, *182*, 107863. [CrossRef]
20. Acosta, M.; Reyes, L.H.; Cruz, J.C.; Pradilla, D. Demulsification of Colombian heavy crude oil (W/O) emulsions: Insights into the instability mechanisms, chemical structure, and performance of different commercial demulsifiers. *Energy Fuels.* **2020**, *34*, 5665–5678. [CrossRef]
21. Zakharova, L.Y.; Vasilieva, E.A.; Mirgorodskaya, A.B.; Zakharov, S.V.; Pavlov, R.V.; Kashapova, N.E.; Gaynanova, G.A. Hydrotropes: Solubilization of Nonpolar Compounds and Modification of Surfactant Solutions. *J. Mol. Liq.* **2023**, *370*, 120923. [CrossRef]

22. Ravera, F.; Dziza, K.; Santini, E.; Cristofolini, L.; Liggieri, L. Emulsification and Emulsion Stability: The Role of the Interfacial Properties. *Adv. Colloid Interface Sci.* **2021**, *288*, 102344. [CrossRef] [PubMed]
23. Low, J.Y.; Khe, C.S.; Usman, F.; Hassan, Y.M.; Lai, C.W.; You, K.Y.; Lim, J.W.; Khoo, K.S. Review on Demulsification Techniques for Oil/Water Emulsion: Comparison of Recyclable and Irretrievable Approaches. *Environ. Res.* **2024**, *243*, 117840. [CrossRef] [PubMed]
24. Zhang, X.; Zhang, L.; Zheng, D.; Xia, Z.; Peng, M.; Sun, D.; Hu, X.; Peng, X. A Natural Monoterpene Enol for Dispersive Liquid-Liquid Microextraction Based on Solidification of Floating Organic Droplets for Determination of Benzophenone Compounds in Water Samples. *Separations* **2023**, *10*, 1. [CrossRef]
25. Mandrah, K.; Satyanarayana, G.N.V.; Roy, S.K. A Dispersive Liquid-Liquid Microextraction Based on Solidification of Floating Organic Droplet Followed by Injector Port Silylation Coupled with Gas Chromatography-Tandem Mass Spectrometry for the Determination of Nine Bisphenols in Bottled Carbonated Beverages. *J. Chromatogr. A* **2017**, *1528*, 10–17.
26. Gałuszka, A.; Migaszewski, Z.M.; Konieczka, P.; Namieśnik, J. Analytical Eco-Scale for Assessing the Greenness of Analytical Procedures. *TrAC Trends Anal. Chem.* **2012**, *37*, 61–72. [CrossRef]
27. Płotka-Wasylka, J. A New Tool for the Evaluation of the Analytical Procedure: Green Analytical Procedure Index. *Talanta* **2018**, *181*, 204–209. [CrossRef]
28. Pena-Pereira, F.; Wojnowski, W.; Tobiszewski, M. AGREE-Analytical GREEnness Metric Approach and Software. *Anal. Chem.* **2020**, *92*, 10076–10082. [CrossRef]
29. Wojnowski, W.; Tobiszewski, M.; Pena-Pereira, F.; Psillakis, E. AGREEprep- Analytical Greenness Metric for Sample Preparation. *TrAC Trends Anal. Chem.* **2022**, *149*, 116553. [CrossRef]
30. González-Martín, R.; Gutiérrez-Serpa, A.; Pino, V.; Sajid, M. A Tool to Assess Analytical Sample Preparation Procedures: Sample Preparation Metric of Sustainability. *J. Chromatogr. A* **2023**, *1707*, 464291. [CrossRef]
31. Balaraman, J.; Md Saleh, N.; Mat Salleh, M.Z.; Yusoff, F.; Asman, S. Deep Eutectic Solvent Coupled with High Performance Liquid Chromatography for Extraction of Bisphenols and Phthalates from Selected Freshwater Source in Selangor, Malaysia. *Sains Malays.* **2023**, *52*, 1469–1483. [CrossRef]
32. Farajzadeh, M.A.; Ghaleh, M.K.; Pezhhanfar, S.; Mogaddam, M.R.A. In-Syringe Homogeneous Liquid-Liquid Microextraction as an Economical, Convenient, and One-Step Analytical Method for the Extraction and Preconcentration of Five Plasticizers and One Antioxidant from Different City Tap Water Samples. *Microchem. J.* **2024**, *204*, 111051. [CrossRef]
33. Niu, R.; Qin, H.; Tao, Y.; Li, L.; Qiao, L. In Situ, Formation of Deep Eutectic Solvents Based Dispersive Liquid-Liquid Microextraction for the Enrichment of Trace Phthalate Esters in Aqueous Samples. *Microchem. J.* **2023**, *189*, 108537. [CrossRef]
34. González-Castro, M.J.; Uribe-Ares, J.; Muniategui-Lorenzo, S.; Beceiro-González, E. Development of a Dispersive Liquid-Liquid Microextraction Method for the Determination of Plastic Additives in Seawater. *Anal. Methods* **2024**, *16*, 1603–1610. [CrossRef]

molecules

MDPI

Article

Unambiguous Determination of Benzo[a]pyrene and Dibenzo[a,l]pyrene in HPLC Fractions via Room-Temperature Fluorescence Excitation–Emission Matrices

George T. Knecht [1], Stephanie D. Nauth [1], Juan C. Gomez Alvarado [1], Anthony M. Santana [1], Hector C. Goicoechea [2,3] and Andres D. Campiglia [1,*]

1 Department of Chemistry, University of Central Florida, Physical Sciences Building, 4111, Orlando, FL 32816, USA; ju916043@ucf.edu (J.C.G.A.)

2 Laboratorio de Desarrollo Analítico y Quimiometría, Catedra de Química Analítica I, Facultad de Bioquímica y Ciencias Biológicas, Universidad Nacional del Litoral, Ciudad Universitaria, Santa Fe 3000, Argentina

3 Consejo Nacional de Investigacioes Científicas y Técnicas (CONICET), Godoy Cruz 2290, Buenos Aires CP C1425FQB, Argentina

* Correspondence: andres.campiglia@ucf.edu

Abstract

When high-performance liquid chromatography (HPLC) is used for the analysis of polycyclic aromatic hydrocarbons (PAHs) in complex samples, further examination of HPLC fractions is recommended to confirm PAH assignments solely based on retention times. Gas chromatography–mass spectrometry (GC-MS) has been particularly relevant in the unambiguous determination of PAHs with remarkably similar retention times. The combination of HPLC and GC requires lengthy analysis times to ensure proper assignments. This article presents an approach for the analysis of co-eluted PAHs with no need for further chromatographic separation. Benzo[a]pyrene (BaP) and dibenzo[a,l]pyrene (DBalP) were directly determined in a co-eluted HPLC fraction via room-temperature fluorescence excitation–emission matrices (RTF-EEMs). RTF-EEMs can be recorded in a matter of seconds with a spectrofluorometer equipped with a multichannel detection system. The spectral overlapping of BaP and DBalP was resolved using parallel factor analysis (PARAFAC). The analytical advantages of this approach were demonstrated with the trace analysis (ng/mL) of these two PAHs in pre-concentrated tobacco extracts.

Keywords: HPLC; room-temperature fluorescence; excitation emission matrices; PARAFAC; benzo[a]pyrene; dibenzo[a,l]pyrene

1. Introduction

HPLC is a popular approach to the qualitative and quantitative analysis of polycyclic aromatic compounds in many scientific fields. Commercial instrumentation for the HPLC analysis of PAHs is usually equipped with ultraviolet–visible (UV-VIS) and/or fluorescence (FL) detectors. Although multichannel detectors with on-the-fly spectral scanning capabilities are intrinsically more selective than single-channel detectors, they lack specificity for the unambiguous identification of co-eluted compounds. This is particularly true for the analysis of PAHs with similar HPLC retention behaviors and, therefore, results in almost identical retention times [1–3]. The unambiguous identification of co-eluted PAHs often requires further analysis of the co-eluted HPLC fractions with a supporting analytical technique such as GC-MS. Unfortunately, the combination of HPLC and GC requires costly experimental procedures with rather lengthy analysis times to ensure proper PAH assignments.

Due to their ubiquitous occurrence and extreme eco-toxicological relevance, the U.S. Environmental Protection Agency (EPA) recommends the routine monitoring of sixteen PAHs with molecular masses (MMs) ranging from 128 Da to 278 Da. These include naphthalene, acenaphthene, acenaphthylene, fluorene, phenanthrene, anthracene, fluoranthene, pyrene, benzo[*a*]anthracene, chrysene, benzo[*b*]fluoranthene, benzo[*k*]fluoranthene, BaP, dibenzo[*a,h*]anthracene, benzo[*g,h,i*]perylene, and indeno[1,2,3-*cd*]pyrene. According to the International Agency for Research on Cancer (IARC), BaP has the highest cancer risk among the EPA PAHs [4].

In addition to the sixteen priority pollutants, the toxicity of PAH-contaminated samples is attributed to the presence of high-molecular-weight PAHs (HMW-PAHs), i.e., PAHs with an MM greater than 300 Da. As a matter of fact, certain PAHs with an MM of 302 Da have shown higher toxicity than the sixteen EPA PAHs. These include DBalP, which is the most potent carcinogenic PAH known to date [5–8]. Its toxicity is approximately 100 times the toxicity of BaP. Thus, unambiguous identification of DBalP in contaminated samples is extremely relevant even if it is present at lower concentrations than other PAHs.

In a recent article, Campiglia and co-workers applied EPA Method 610 to the HPLC analysis of 15 EPA PAHs and 10 HMW-PAHs with an MM of 302 Da from tobacco [9]. Except for DBalP, all the 302 Da isomers showed the expected chromatographic behaviors under reversed-phase conditions, i.e., they eluted later than the EPA PAHs from the chromatographic column due to the stronger affinities for the octadecyl stationary phase [2]. DBalP co-eluted with BaP and therefore, eluted earlier than three heaviest EPA PAHs, namely dibenzo[*a,h*]anthracene, benzo[*g,h,i*]perylene, and indeno[1,2,3-*cd*]pyrene.

The unambiguous determination of DBalP and BaP in HPLC fractions was then accomplished via Laser-Excited Time-Resolved Shpol'skii Spectroscopy (LETRSS) [9]. The two PAHs were collected within their time-window of HPLC elution (29-32 min). The mobile phase (acetonitrile) was evaporated to dryness and the residue was reconstituted in 1mL of octane. The LETRSS analysis was carried out at 77 K with the aid of a cryogenic fiber optic probe and an instrumental set-up that was built in-house for time-resolved laser-induced fluorescence measurements. Wavelength time matrices (WTMs) were collected from the frozen samples to obtain line-narrowed spectra and fluorescence lifetimes resulting in the unambiguous determination of BaP and DBalP.

The present study proposes a straightforward approach to the analysis of these two PAHs in HPLC fractions. It consists of pouring the HPLC fraction directly into a quartz cuvette to record RTF-EEMs using a commercial spectrofluorometer. Since the spectrofluorometer is equipped with a spectrograph and a charge-coupled device (CCD), recording RT-EEMs from HPLC fractions only takes a few seconds per sample. Spectral deconvolution was accomplished with PARAFAC. Although this algorithm has been applied to the analysis of several EPA PAHs via RTF [10–17], our literature search did not find any reports of the application of PARAFAC for the unambiguous determination of BaP and DBalP in HPLC fractions. The same was true for the analysis of these two PAHs in tobacco samples. Although B[a]P has been found in a variety of tobacco products and cigarette smoke, there are no reports that detected the presence of DBalP using chromatographic methods [18–21]. This is rather intriguing, particularly if one considers the incomplete combustion of tobacco during the smoking process. The straightforward experimental procedure, the excellent analytical figures of merit, and the accuracy of the analysis presented here make this approach a robust alternative for the routine analysis of BaP and DBalP in tobacco samples.

2. Results and Discussion

Figure 1 shows a typical chromatogram of a standard mixture containing the 25 studied PAHs [9]. The co-elution of BaP and DBalP occurred between an elution time of approximately 30 and 32 min, and it was further confirmed by the statistical equivalence ($P \geq 98\%$; $n_1 = n_2 = 3$) [22] of their retention times ($t_R^{BaP} = 30.9 \pm 0.3$ min; $t_R^{DBalP} = 31.8 \pm 0.3$ min). Considering the experimental simplicity of pouring the HPLC fraction into a cuvette for measurements using a spectrofluorometer, we studied the RTF features of BaP and DBalP in acetonitrile, i.e., the mobile phase of the HPLC separation.

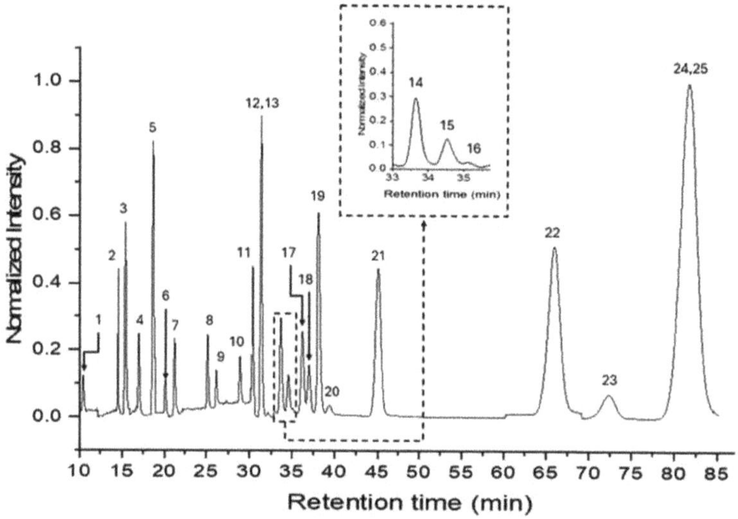

Figure 1. Typical HPLC chromatogram recorded from a standard solution containing the 16 EPA PAHs and 10 HMW-PAHs in acetonitrile. All PAH concentrations were at the ng/mL level. Fluorescence measurements were performed at the maximum excitation and emission wavelengths of each PAH [9].

2.1. Room Temperature Excitation and Fluorescence Spectra of BaP and DBalP

Figure 2A and 2B show the excitation and emission spectra of BaP and DBalP recorded from pure standard solutions in acetonitrile, respectively. All spectra were recorded at the maximum excitation and emission wavelengths of each PAH using a 3 nm band-pass for both monochromators. Although the two PAHs showed characteristic spectral profiles with distinct excitation and emission wavelengths, the strong spectral overlapping observed in Figure 2C might prevent their unambiguous determination in HPLC fractions. This possibility was investigated with binary mixtures of BaP and DBalP in acetonitrile. Table S1 summarizes the results obtained for BaP. The possible interference of DBalP was tested at the same concentration of BaP (1:1 mixture) and at $5 \times$ (1:5 mixture) and $10 \times$ (1:10 mixture) higher concentrations of DBalP compared to BaP. All fluorescence intensities were measured at the maximum excitation and emission wavelength of BaP and then compared to the fluorescence intensity of a pure standard solution of BaP. No interference from DBalP was observed for the 1:1 mixture. As shown in Table S2, similar results were obtained for DBalP in the presence of BaP.

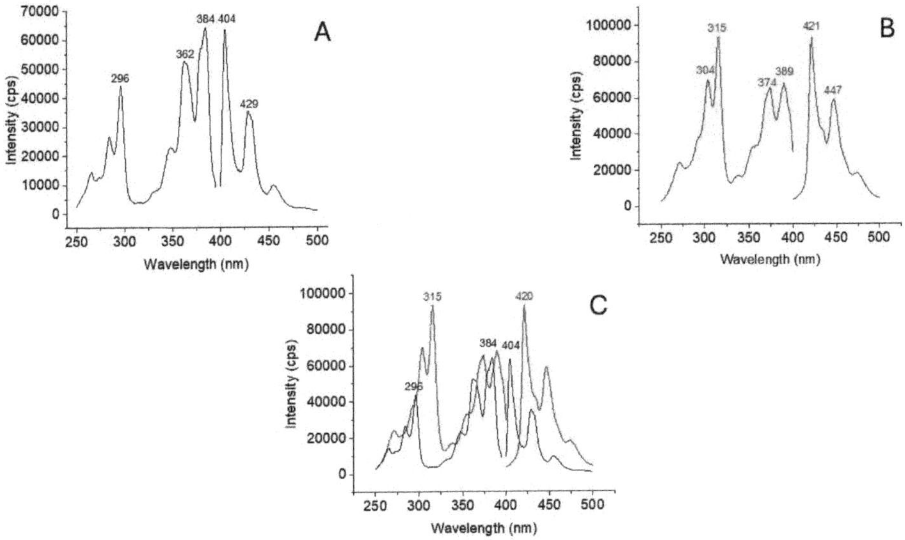

Figure 2. Excitation and fluorescence spectra of BaP (**A**) and DBalP (**B**) recorded from pure standard solutions in acetonitrile. The excitation and emission spectra were recorded at the maximum emission and excitation wavelengths of each PAH, respectively. The excitation and emission passbands were 3 nm in all cases. Figure (**C**) illustrates the spectral overlap of these two PAHs in acetonitrile solutions. All spectra were normalized for the maximum fluorescence intensities.

2.2. Room-Temperature Fluorescence Excitation–Emission Matrices (RTF-EEMs) of BaP and DBalP

The unambiguous determination of co-eluted PAHs in HPLC fractions is limited by the broad nature of room-temperature excitation and fluorescence spectra. Well-known approaches to reducing the overlap of excitation and fluorescence spectra include coupling high-order instrumental data to parallel factor analysis (PARAFAC). This approach carries with it the second-order advantage, which permits the quantification of fluorophores in samples of unknown composition no matter how many signal-overlapping constituents are in the unknown sample [23–27].

Previous work in Campiglia's lab on the analysis of EPA PAHs in soil samples combined PARAFAC with 4.2K time-resolved EEMs (4.2K TREEMs) [28]. TREEMs refer to excitation–emission matrices recorded with laser-based instrumentation at certain time windows during the total fluorescence decay of the sample. The present study combined PARAFAC and RT-EEMs, which consisted of fluorescence spectra recorded at several excitation wavelengths and compiled into either a two-dimensional data format (wavelength of excitation vs. wavelength of fluorescence) or a three-dimensional format (wavelength of excitation vs. wavelength of fluorescence vs. signal intensity).

Figure 3A,B show the RT-EEMs recorded from pure standard solutions of BaP and DBalP in acetonitrile. Figure 3C shows the RTF-EEM recorded from a binary mixture of the two PAHs in acetonitrile. The excitation and emission passbands were set at 3 nm in all cases. The sample excitation was conducted at 3 nm increments between 250 and 393 nm. The excitation wavelengths were scanned from low (393 nm) to high energy (250 nm) to reduce the sample exposure to UV irradiation and possible photobleaching. The fluorescence emission was monitored between 394 and 502 nm using an integration time of 150 milliseconds. Scatter interference from excitation radiation and from second-order grating effects were eliminated with the aid of fully automated cut-off filters. Under these instrumental settings, each EEM resulted in a data matrix of 55 excitation data points $\times$ 48 emission data points, with an average acquisition time of 20.7 $\pm$ 0.3 s per EEM.

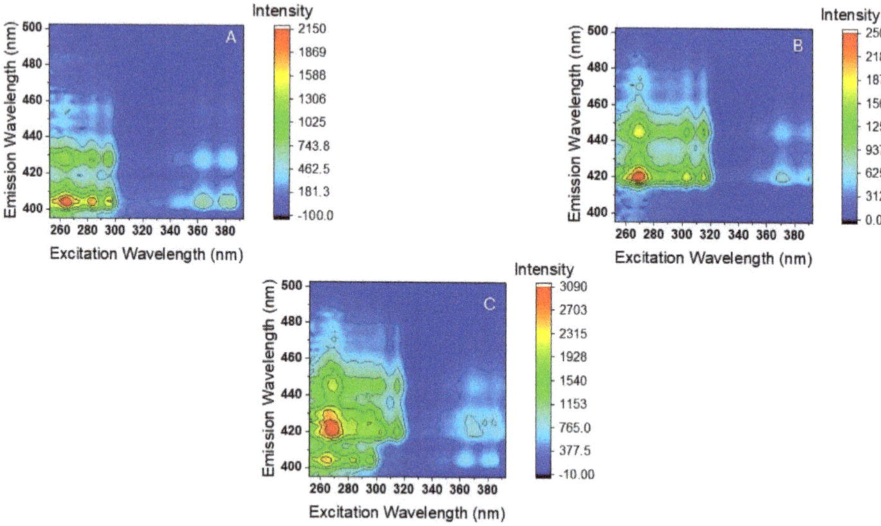

Figure 3. RTF-EEMs recorded from pure standard solutions of (**A**) BaP and (**B**) DBalP in acetonitrile and (**C**) a binary mixture of BaP and DBalP in acetonitrile. In all cases, the concentrations of BaP and DBalP were 29 ng/mL and 281 ng/mL, respectively. All measurements were performed using 3 nm excitation and emission passbands.

2.3. *RTF-EEMs' Analytical Figures of Merit for the Analysis of BaP and DBalP in Acetonitrile*

The analytical figures of merit (AFOM) for the analysis of BaP and DBalP in acetonitrile are summarized in Table 1. The fluorescence intensities plotted in the calibration graphs correspond to the intensity values obtained from the EEMs at the maximum excitation and emission wavelengths of each PAH. No attempts were made to experimentally determine the upper concentration limits of the linear dynamic ranges (LDRs). The correlation coefficients (Rs) were close to unity, demonstrating linear correlations between fluorescence intensity and PAH concentration. The LOD and the LOQ of BaP were approximately one order of magnitude lower than those of DBalP. These values are in good agreement with the stronger fluorescence emitted by BaP. The relative standard deviations (RSDs) at medium linear concentrations demonstrated good reproducibility at the parts-per-billion (ng/mL) concentration level.

Table 1. RTF-EEMs' analytical figures of merit for BaP and DBalP in acetonitrile.

	$\lambda_{exc}/\lambda_{em}$ [1]	LOD [2] (ng/mL)	LOQ [3] (ng/mL)	LDR [4] (ng/mL)	R [5]	% RSD [6]
BaP	384/404	4.7	16	16–104	0.9995	0.6
DBalP	315/421	32	107	107–500	0.9988	4.2

[1] Excitation (λ_{exc}) and emission (λ_{em}) wavelengths used for fluorescence measurements. [2] Limit of detection (LOD) was calculated using LOD = $3S_B/m$, where S_B is the standard deviation of sixteen blank signal measurements and m is the slope of the calibration curve. [3] Limit of quantitation (LOQ) was calculated using LOQ = $10S_B/m$. [4] Linear Dynamic Range (LDR), in ng/mL, extends from the LOQ to an arbitrarily chosen upper linear concentration. [5] Correlation coefficient (R) of LDR. [6] Relative standard deviation (RSD) = S/I $\times$ 100, where I is the average intensity and S is the standard deviation of the intensity calculated from three measurements at medium linear concentrations.

2.4. Calibration and Validation Sets for PARAFAC Analysis

If EEMs are arranged in a three-dimensional representation $\mathbf{X}$ of dimensions $I \times J \times K$, where I is the number of samples, J is the number of emission wavelengths, and K is the number of excitation wavelengths, PARAFAC can decompose the array into a more condensed form than the original one [23]. This is accomplished by minimizing the sum of the squares of the residuals (e_{ijk}) in the model described by Equation (1):

$$x_{ijk} = \sum_{n=1}^{n} a_{in} b_{jn} c_{kn} + e_{ijk} \quad (1)$$

where n indicates the component number and x_{ijk} is the fluorescence intensity for sample i at the emission wavelength j and excitation wavelength k. For any given component n, the elements a_{in}, b_{jn}, and c_{kn} are arranged in the score vector $\mathbf{a}_n$ (whose elements are directly proportional to its concentration in each sample) and the loading vectors $\mathbf{b}_n$ and $\mathbf{c}_n$, which estimate its emission and excitation profiles.

Binary mixtures with different concentrations of BaP and DBalP were prepared in acetonitrile for the calibration and the validation sets. Three individual EEMs were recorded from three aliquots of each binary mixture to account for possible instrumental variations. The same approach was used to record EEMs from the blank solution (acetonitrile). The average plots of the blank-subtracted EEMs recorded from all the binary mixtures in the calibration set and the validation set are shown in Figure S1 and Figure S2 respectively.

Table 2 summarizes the composition of the calibration set for the PARAFAC model employed in these studies. All solutions consisted of standard mixtures of BaP and DBalP in acetonitrile. The concentrations of the two PAHs were calculated with a Central Composite Design (CCD) and took into consideration the LDRs in Table 1.

Table 2. Concentrations of BaP and DBalP in binary mixtures employed for PARAFAC calibration.

Sample	BaP (ng/mL)	DBalP (ng/mL)
1	28.9	164.6
2	91.1	164.6
3	28.9	442.4
4	91.1	442.4
5	16.0	303.5
6	104.0	303.5
7	60.0	107.0
8	60.0	500.0
9	60.0	303.5
10	0.0	303.5
11	60.0	0.0

Because BaP is a stronger fluorophore than DBalP, its concentration in all the binary mixtures was considerably lower than the concentration of DBalP. In addition, a zero-concentration sample for each analyte was included in the calibration sets. Table 3 lists the nominal concentrations of BaP and DBalP in the binary mixtures for the validation set and those predicted by PARAFAC using two factors. For the binary mixtures, there was no need to exploit the second-order advantage. The nominal concentrations of the standard mixtures were different from those in Table 1 but within the same concentration ranges as the calibration set. A comparison of the nominal to the predicted concentration clearly showed an excellent predictive ability, with mean recoveries of 102.2 ± 5% for BaP and 100.4 ± 3% for DBalP.

Table 3. Concentrations of BaP and DBalP in binary mixtures employed for PARAFAC validation.

Sample	BaP (ng/mL)		DBalP (ng/mL)	
	Nominal [1]	Predicted [2]	Nominal [1]	Predicted [2]
1	39.5	37.8	159.8	165.0
2	85.5	87.9	159.8	156.9
3	39.5	40.9	400.2	382.9
4	85.5	89.9	400.2	431.5
5	30.0	32.6	280.0	267.2
6	95.0	87.9	280.0	277.2
7	62.5	64.1	110.0	117.2
8	62.5	65.7	450.0	428.4
9	62.5	64.3	280.0	286.9

[1] Nominal concentration refer to the actual concentrations of BaP and DBalP in the binary mixture. [2] Predicted concentrations refer to the concentrations calculated by PARAFAC.

2.5. HPLC–EEM Analysis of Tobacco Smoke Condensate

Five samples of tobacco smoke condensate (TSC) were subjected to HPLC analysis. Four samples were prepared by mixing aliquots of pre-concentrated TSC to solid amounts of BaP and DBalP. The remaining sample consisted of TSC with no addition of BaP and DBalP. Figure 4 shows the RTF-EEMs recorded from the five HPLC samples.

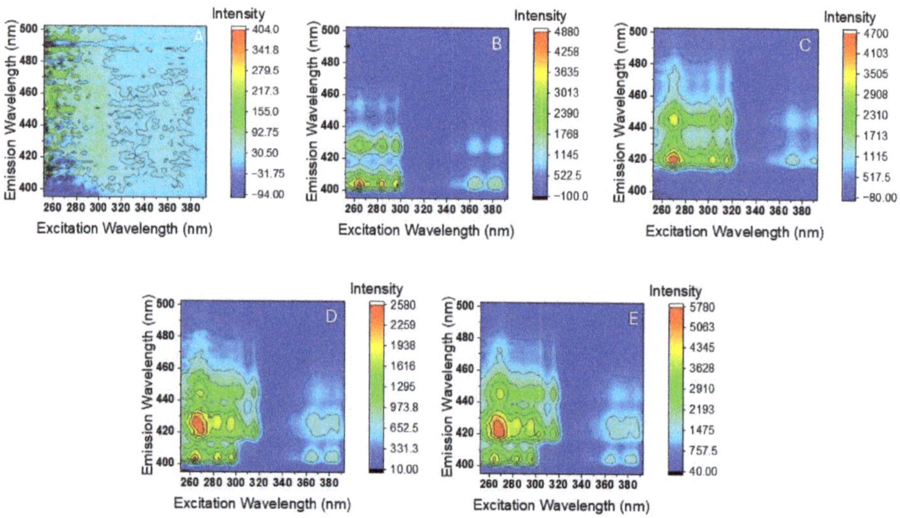

Figure 4. RTF-EEMs of HPLC fractions recorded from TSC mixed with different concentrations of BaP and DBalP (**B–E**) and from unmixed TSC (**A**).

A visual comparison of the EEMs in Figure 3 provides insights into the contribution of each PAH to the total fluorescence of the HPLC fractions. The nominal concentrations of BaP and DBalP in the five TSC samples are listed in Table 4. It should be noted that the reported concentrations only compute the amounts of BaP and DBalP added to the TSC samples. The excellent agreement between the nominal and the predicted concentration demonstrate the ability of PARAFAC to determine the two PAHs in co-eluted HPLC fractions. These quantitative results are consistent with the quality of the profiles extracted by PARAFAC for both analytes (see SI Figure 3).

Table 4. PARAFAC predictions of BaP and DBalP concentrations spiked into samples of tobacco.

Sample	BaP (ng/mL)		DBalP (ng/mL)	
	Nominal [1]	Predicted [2]	Nominal [1]	Predicted [2]
1	0.0	–	0.0	–
2	85.5	88.4	0.0	–
3	0.0	–	400.2	387.2
4	39.5	38.3	159.8	162.1
5	85.5	87.9	400.2	406.1
Average Recovery (%)	–	100.9	–	99.9

[1] Nominal concentration refer to the actual concentrations of BaP and DBalP in the binary mixture. [2] Predicted concentrations refer to the concentrations calculated by PARAFAC.

3. Materials and Methods

3.1. Chemicals

All reagents used were the highest available purity. Analytical standards of BaP and DBalP were purchased from Chiron (Trondheim, Norway). Pure (100%) HPLC-grade acetonitrile and acetone were purchased from Fisher Scientific (Waltham, MA, USA). Camel Menthol Crush commercial cigarettes were purchased at local stores. Nanopure water supplied by the Barnstead Nanopure Infinity water system (Barnstead, Dubuque, IA, USA) located in the laboratory was used throughout all the experiments.

3.2. Preparation of Calibration and Validation Sets for PARAFAC Analysis

Calibration and validation samples were prepared by diluting stock solutions of BaP and DBalP in HPLC-grade acetonitrile. All dilutions were made with micropipettes to a total volume of 3.00 mL. The PAH final concentrations varied between 16.0 ng/mL and 104.0 ng/mL for BaP, and between 164.6 ng/mL and 500.0 ng/mL for DBalP. In addition, a zero-concentration sample for each analyte was included in the calibration sets. For the validation sets, nine solutions were prepared with concentrations varying from 30.0 ng/mL to 95.0 ng/mL for BaP and from 110 ng/mL to 450 ng/mL for DBalP.

3.3. Collection of Tobacco Smoke Condensate

Tobacco smoke condensate was collected according to the guidelines of the National Institute of Standards and Technology (NIST SRM 3222) [29]. The main steps were as follows: A sintered glass frit funnel was placed on an Erlenmeyer vacuum flask with a length of tubing extending from the funnel into 300 mL of acetone. The apparatus was placed in dry ice and a slight vacuum was created. After setup, 11 g of cigarette tobacco filler was ignited and burned to ash in the filter, such that the smoke bubbled through the acetone. After 2 min, the ashes were stoked, and the apparatus allowed to cool for 8 min. After cooling, the remaining ashes were removed, and the funnel was washed with around 200 mL of acetone. The funnel and tubing were then removed and sonicated in the extract solution to remove any remaining condensate.

3.4. HPLC–EEM Analysis of Tobacco Smoke Condensate

Prior to HPLC analysis, a pre-concentration step for the tobacco smoke condensate was performed by evaporating 8 mL of the condensate to a residue, which was then reconstituted with acetonitrile to a final volume of 1.0 mL and centrifuged for 1 min at 10,000 rpm. Samples for PARAFAC analysis were prepared by adding 100 µL aliquots of the pre-concentrated smoke condensate to solid residues containing various masses of either BaP or DBalP. BaP residues were prepared by evaporating different volumes of a 101.2 mg/mL BaP standard solution in acetonitrile. DBalP residues were prepared by evaporating different volumes of a 513.0 mg/mL DBalP standard solution in acetonitrile. All mixtures were centrifuged at 600 rpm for 5 min to ensure complete dissolution of the solid residue into the liquid aliquot of the pre-concentrated smoke condensate. All sample injections were performed using a 20 μL fixed-volume injection loop. HPLC fractions were collected from 27.5 to 31.5 min of the chromatographic time and subjected to EEM analysis. Each HPLC fraction was subjected to triplicate measurements made from three sample injections.

3.5. Instrumentation

Chromatographic analysis was performed with a Hitachi HPLC system (San Jose, CA, USA) using a computer with Hitachi software. Its main components included a model L7100 gradient pump, an L-7400 UV detector, an L-7485 fluorescence detector, an L-761 online degasser, and a D-7000 control interface. PAH separation was accomplished with an Eclipse PAH column (250 mm length, 4.6 mm inner diameter, and 5 μm particle size), using isocratic elution for five minutes with 40% acetonitrile and 60% water, and then linear gradient elution to 100% with acetonitrile over 25 min. The mobile phase flow rate was 2.0 mL/min and the column temperature was ~25 °C. Sample injection (20 μL) was performed using a fixed-volume injection loop. HPLC fractions were collected in 30 mL amber vials.

Room-temperature fluorescence measurements were made using a Duetta™ spectrofluorometer (Horiba Scientific, Piscataway, NJ, USA) equipped with a 75 W xenon arc lamp, a single-grating scanning monochromator for sample excitation, and a spectrograph/TE-cooled CCD for spectra collection from 250 nm to 1100 nm. The instrument is equipped with a reference detector (Si photodiode) to compensate for variations in the intensity of sample excitation and a fully automated set of cut-off filters to eliminate second-order grating effects. Instrument control was accomplished with custom software (EzSpec™ Touch-Screen Software, https://www.horiba.com/jpn/scientific/products/detail/action/show/Product/ezspec-software-1880/#show-more). All measurements were made from liquid solutions in 500 μL quartz cuvettes.

3.6. Software for Data Analysis with PARAFAC

MATLAB 7.10 was used for all calculations (The Mathworks Inc., Natick, MA, USA, 2010) [30]. Software processing was facilitated by an MVC2 graphical user interface written in MATLAB that was previously described by Olivieri et al. [31].

4. Conclusions

Although DBalP is the most toxic PAH known to date, there is relatively limited information on its possible presence in tobacco samples. Our study showed that DBalP co-elutes with BaP under standard HPLC conditions. Therefore, their unambiguous identification requires further analysis of the HPLC fractions via an alternative analytical method. One possibility is to perform the analysis via GC-MS. This technique is able to differentiate BaP from DBalP in HPLC fractions, but it would require further separation via a rather lengthy experimental procedure.

Herein, an alternative method was proposed to obviate further chromatographic separation of these two PAHs in co-eluted HPLC fractions. The experimental procedure is straightforward; it consists of pouring the HPLC fraction directly into a quartz cuvette to record RTF-EEMs using a commercial spectrofluorometer. If the spectrofluorometer is equipped with a multichannel detection system, i.e., a spectrograph and a CCD, the RTF-EEM recording will take less than one minute per HPLC fraction. These are distinct advantages compared to LETRSS analysis [9], which requires laser-based instrumentation and the use of a liquid cryogen to avoid spectral overlap. In the present study, the strong spectral overlap between BaP and DBalP was resolved at room temperature with the aid of PARAFAC. Since this algorithm carries with it the second-order advantage, it makes recording RTF-EEMs an accurate approach for the unambiguous determination of BaP and DBalP at the parts-per-billion level via EPA Method 610.

Supplementary Materials: The following supporting information can be downloaded at: https://www.mdpi.com/article/10.3390/molecules30071550/s1, Table S1: Statistical comparison of fluorescence intensities recorded at the maximum excitation and emission wavelengths of BaP in 1:1, 1:5, and 1:10 binary mixtures with DBalP in acetonitrile; Table S2: Statistical comparison of fluorescence intensities recorded at the maximum excitation and emission wavelengths of DBalP in 1:1, 1:5, and 1:10 binary mixtures with BaP in acetonitrile; Figure S1: RTF-EEMs recorded from the calibration set used for the PARAFAC analysis; Figure S2: RTF-EEMs recorded from the validation set used for the PARAFAC analysis; Figure S3: Examples of spectral profiles extracted by PARAFAC from HPLC fractions.

Author Contributions: G.T.K.: collection of room-temperature fluorescence excitation–emission matrices; S.D.N. and J.C.G.A.: HPLC analysis; A.M.S.: pre-concentration of tobacco smoke condensate with and without BaP and DBalP; H.C.G.: PARAFAC analysis; A.D.C.: formulation of research goals and aims; development of methodology; management, planning, and execution of research activity; review and editing. All authors have read and agreed to the published version of the manuscript.

Funding: This research received no external funding.

Institutional Review Board Statement: Not applicable.

Informed Consent Statement: Not applicable.

Data Availability Statement: Data are contained within the article and Supplementary Materials.

Conflicts of Interest: The authors declare no conflicts of interest.

References

1. Roberts, K.P.; Jankowiak, R.; Small, G.J. High-performance liquid chromatography interfaced with fluorescence line-narrowing spectroscopy for on-line analysis. *Anal. Chem.* **2001**, *73*, 951–956. [CrossRef] [PubMed]
2. Wise, S.A.; Rodgers, R.P.; Reddy, C.M.; Nelson, R.K.; Kujawinski, E.B.; Wade, T.L.; Campiglia, A.D.; Lieu, Z. Advances in Chemical Analysis of Oil Spills Since the *Deepwater Horizon* Disaster. *Crit. Rev. Anal. Chem.* **2022**, *53*, 1638–1697. [CrossRef] [PubMed]
3. Tian, B.Y.; Gao, S.T.; Zhu, Z.J.; Zeng, X.Y.; Liang, Y.; Yu, Z.Q.; Peng, P.A. Two-dimensional gas chromatography coupled to isotope ratio mass spectrometry for determining high molecular weight polycyclic aromatic hydrocarbons in sediments. *J. Chromatogr. A* **2023**, *1693*, 463879. [CrossRef]
4. IARC Working Group on the Evaluation of Carcinogenic Risks to Humans Chemical Agents and Related Occupations. Lyon (FR): International Agency for Research on Cancer. IARC Monographs on the Evaluation of Carcinogenic Risks to Humans, No. 100F BENZO[a]PYRENE, 2012. Available online: https://www.ncbi.nlm.nih.gov/books/NBK304415/ (accessed on 14 March 2025).
5. Cavalieri, E.L.; Higginbotham, S.; Rogan, E.G. Dibenzo[a,l]pyrene—The Most Potent Carcinogenic Aromatic Hydrocarbon. *Polycycl. Aromat. Compd.* **1994**, *6*, 177–183. [CrossRef]
6. Mahadevan, B.; Luch, A.; Bravo, C.F.; Atkin, J.; Steppan, L.B.; Pereira, C.; Kerkvliet, N.I.; Baird, W.M. Dibenzo[a,l]pyrene induced DNA adduct formation in lung tissue in vivo. *Cancer Lett.* **2005**, *227*, 25–32. [CrossRef]
7. Russell, G.K.; Gupta, R.C.; Vadhanam, M.V. Effect of phytochemical intervention on dibenzo[a,l]pyrene-induced DNA adduct formation. *Mutat. Res./Fundam. Mol. Mech. Mutagen.* **2015**, *774*, 25–32. [CrossRef]
8. Kowalczyk, K.; Roszak, J.; Sobanska, Z.; Stepnik, M. Review of mechanisms of genotoxic action of dibenzo[def,p]chrysene (formerly dibenzo[a,l]pyrene). *Toxin Rev.* **2023**, *42*, 460–477. [CrossRef]
9. Comas, A.; Santana, A.; Campiglia, A.D. On the co-elution of benzo[a]pyrene and dibenzo[a,l]pyrene in chromatographic fractions and their unambiguous determination in tobacco extracts via laser-excited time resolved Shpol'skii spectroscopy. *Anal. Methods* **2023**, *15*, 1959–1968. [CrossRef]
10. Elcoroaristizabal, S.; de Juan, A.; Garcia, A.J.; Durana, N.; Alonso, L. Comparison of second-order multivariate methods for screening and determination of PAHs by total fluorescence spectroscopy. *Chemom. Intell. Lab. Syst.* **2014**, *132*, 63–64. [CrossRef]

11. Ahmadvand, M.; Sereshti, H.; Parastar, H. Chemometric-based determination of polycyclic aromatic hydrocarbons in aqueous samples using ultrasound-assisted emulsification microextraction combined to gas chromatography–mass spectrometry. *J. Chromatogr. A* **2015**, *1413*, 117–126. [CrossRef]
12. Gu, H.W.; Zhang, S.H.; Wu, B.C.; Chen, W.; Wang, J.B.; Liu, Y. A green chemometrics-assisted fluorimetric detection method for the direct and simultaneous determination of six polycyclic aromatic hydrocarbons in oil-field wastewaters. *Spectrochim. Acta A Mol. Biomol. Spectrosc.* **2018**, *200*, 93–101. [CrossRef] [PubMed]
13. Panigrahi, S.K.; Mishra, A.K. Identification of pyrene in complex sample matrix using time-resolved fluorescence measurement coupled with PARAFAC analysis. *J. Photochem. Photobiol. A* **2019**, *383*, 111991. [CrossRef]
14. Catena, S.; Sanllorente, S.; Sarabia, L.A.; Boggia, R.; Turrini, F.; Ortiz, M.C. Unequivocal identification and quantification of PAHs content in ternary synthetic mixtures and in smoked tuna by means of excitation-emission fluorescence spectroscopy coupled with PARAFAC. *Microchem. J.* **2020**, *154*, 10456. [CrossRef]
15. Araújo, K.C.; Barreto, M.C.; Siqueira, A.S.; Freitas AC, P.; Oliveira, L.G.; Bastos ME, P.A.; Rocha ME, P.; Silva, L.A.; Fragoso, W.D. Oil spill in northeastern Brazil: Application of fluorescence spectroscopy and PARAFAC in the analysis of oil-related compounds. *Chemosphere* **2021**, *267*, 129154. [CrossRef]
16. Seopela, M.P.; Powers, L.C.; Clark, C.; Heyes, A.; Gonsior, M. Combined fluorescent measurements, parallel factor analysis and GC-mass spectrometry in evaluating the photodegradation of PAHS in freshwater systems. *Chemosphere* **2021**, *269*, 129386. [CrossRef] [PubMed]
17. Siqueira, A.S.; Almeida, L.F.; Fragoso, W.D. Determination of anthracene, phenanthrene, and fluorene in tap water and sediment samples by fluorescence spectroscopy on nylon membranes and second-order calibration. *Talanta* **2023**, *253*, 124002. [CrossRef]
18. Hearn, B.A.; Ding, Y.S.; Watson, C.H.; Johnson, T.L.; Zewdie, G.; Jeong-Im, J.H.; Walters, M.J.; Holman, M.R.; Rochester, C.G. Multi-year Study of PAHs in Mainstream Cigarette Smoke. *Tob. Regul. Sci.* **2018**, *4*, 96–106. [CrossRef] [PubMed]
19. Palazzi, P.; Hardy, E.M.; Appenzeller, B.M.R. Biomonitoring of children exposure to urban pollution and environmental tobacco smoke with hair analysis—A pilot study on children living in Paris and Yeu Island, France. *Sci. Total Environ.* **2019**, *665*, 864–872. [CrossRef]
20. Adesina, O.A.; Olowolafe, T.I.; Igbafe, A. Levels of polycyclic aromatic hydrocarbon from mainstream smoke of tobacco products and its risks assessment. *J. Hazard. Mater. Adv.* **2022**, *5*, 100053. [CrossRef]
21. Burkhardt, T.; Scherer, M.; Scherer, G.; Pluym, N.; Weber, T.; Kolossa-Gehring, M. Time trend of exposure to secondhand tobacco smoke and polycyclic aromatic hydrocarbons between 1995 and 2019 in Germany—Showcases for successful European legislation. *Environ. Res.* **2023**, *216 Pt 2*, 114638. [CrossRef]
22. Miller, J.N.; Miller, J.C.; Miller, R.D. *Statistics and Chemometrics for Analytical Chemistry*, 7th ed.; Pearson: London, UK, 2018.
23. Murphy, K.R.; Stedmon, C.A.; Graeber, D.; Bro, R. Fluorescence spectroscopy and multi-way techniques. PARAFAC. *Anal. Methods* **2013**, *5*, 6557–6566. [CrossRef]
24. Zhang, W.; Li, T.; Dong, B. Characterizing dissolved organic matter in Taihu Lake with PARAFAC and SOM method. *Water Sci. Technol.* **2022**, *85*, 706–718. [CrossRef]
25. Dinç, E.; Ertekin, Z.C.; Ünal, N. Three-way analysis of pH-UV absorbance dataset for the determination of paracetamol and its pKa value in presence of excipients. *Spectrochim. Acta Part A* **2020**, *230*, 118049. [CrossRef] [PubMed]
26. Olivieri, A. Analytical Figures of Merit in Univariate, Multivariate, and Multiway Calibration: What Have We Learned? What Do We Still Need to Learn? *J. Chemom.* **2024**, *38*, e3613. [CrossRef]
27. Olivieri, A.C.; Escandar, G.M. Recent Advances in Multiway Analytical Figures of Merit. *Data Handl. Sci. Technol.* **2024**, *33*, 363–380. [CrossRef]
28. Goicoechea, H.C.; Yu, S.; Moore AF, T.; Campiglia, A.D. Four-way modeling of 4.2 K time-resolved excitation emission fluorescence data for the quantitation of polycyclic aromatic hydrocarbons in soil samples. *Talanta* **2012**, *101*, 330–336. [CrossRef] [PubMed]
29. *SRM 3222*; Collection of Tobacco Smoke Condensate (SRM 3222 Cigarette Tobacco Filler). National Standards of Technology. Chemical Sciences Division, Material Measurement Laboratory, NIST: Gaithersburg, MD, USA, 2017.
30. Bro, R. PARAFAC Tutorial and applications. *Chemometr. Intell. Lab. Syst.* **1997**, *38*, 149–171. [CrossRef]
31. Chiappini, F.A.; Munoz de la Pena, A.; Goicoechea, H.C.; Olivieri, A.C. An upgrade of MVC2, a MATLAB graphical user interface for second-order multivariate calibration: Beyond trilinear models. *Chemometr. Intell. Lab. Syst.* **2023**, *237*, 104814. [CrossRef]

molecules

MDPI

Article

Green Analytical Method Using Single-Drop Microextraction Followed by Gas Chromatography for Nitro Compound Detection in Environmental Water and Forensic Rinse Water

Tamara Pócsová, Senad Okanovič and Svetlana Hrouzková *

Faculty of Chemical and Food Technology, Institute of Analytical Chemistry, Slovak University of Technology in Bratislava, Radlinského 9, 812 37 Bratislava, Slovakia; tamara.pocsova@stuba.sk (T.P.)
* Correspondence: svetlana.hrouzkova@stuba.sk

Abstract

The extensive use of nitro compounds in agriculture, industry, armaments, and pharmaceuticals, along with their toxic effects on living organisms, necessitates efficient and environmentally sustainable analytical methods. Traditional extraction techniques often involve practices that are not eco-friendly, such as the use of large volumes of solvents, toxic chemicals, and the generation of significant waste; therefore, the single-drop microextraction technique was involved in overcoming these limitations. This study shows an environmentally friendly method for nitro compound analysis focusing on NB (Nitrobenzene), 2-NT (2-Nitrotoluene), 3-NT (3-Nitrotoluene), 4-NT (4-Nitrotoluene), 1,3-DNB (1,3-Dinitrobenzene), 1,2-DNB (1,2-Dinitrobenzene), 2,4-DNT (2,4-Dinitrotoluene), and TNT (Trinitrotoluene). To separate and to detect selected nitro compounds, gas chromatography with an electron capture detector was utilized, which is highly selective for analytes containing nitro groups. To determine optimal experimental conditions, extraction parameters were studied, including the impact of salt addition, temperature, and pH on extraction efficiency. Key performance parameters, such as limit of detection (LOD), limit of quantification (LOQ), repeatability, extraction recoveries, calibration range, and matrix effects, were assessed. The LOD values ranged from 0.01 to 0.09 μg/L in deionized water, 0.01 to 0.06 μg/L in tap water, 0.01 to 0.03 μg/L in seawater, and 0.03 to 0.11 μg/L in model forensic rinse water. The optimized method was successfully applied to the determination of nitro compounds in real environmental water samples and forensic rinse water samples. The environmental sustainability and greenness of the proposed method was evaluated with the AGREE, AGREEprep, and AESA techniques.

Keywords: nitro compounds; forensic samples; environmental water; single-drop microextraction; gas chromatography

1. Introduction

Nitroaromatic compounds are among the largest and most important groups of industrial chemicals currently in use. These compounds are organic molecules that consist of at least one nitro group (-NO_2) attached to an aromatic ring. The vast majority of these are synthetic, although a few biologically produced nitroaromatic compounds have been identified. Nitroaromatic compounds can occur naturally in both atmospheric and aqueous environments. In the urban environment, hydrocarbons released from natural combustion processes and the incomplete combustion of fossil fuels serve as substrates for nitration with atmospheric nitrogen dioxide [1]. In the aquatic environment, solar radiation catalyzes the nitration and halogenation reactions of naturally occurring or anthropogenic compounds in a similar manner [2]. Although the vast majority of nitroaromatic compounds are synthesized chemicals, they have also been discovered as natural products from various bacteria, fungi, and plants [3]. Although nitroaromatic compounds that naturally occur in the environment are relatively rare, they have mainly entered the environment through human activity. This important class of industrial chemicals is used in large volumes in the synthesis of a variety of products including explosives, dyes, polymers, and pesticides. Unfortunately, their widespread use has led to the contamination of soil and groundwater in the environment. The nitro group, which provides the chemical and functional diversity of these molecules, also contributes to the resistance of these compounds to biodegradation. Nitroaromatic compounds are dangerous to human health due to their toxicity, mutagenicity, and easy reduction to carcinogenic aromatic amines [1]. Although some nitroaromatic compounds are intentionally applied to the environment (pesticides), improper handling and/or storage practices by manufacturers and users have led to their accidental release into the environment in countries around the world. The annual tonnage of released chemicals reflects the enormous scale of this problem [4]. Industrial accidents have also led to the contamination of the environment with nitroaromatic compounds. The most significant contamination in recent history occurred on 4 August 2020 in Beirut, Lebanon, where an ammonium nitrate warehouse exploded [5]. Another major environmental contamination occurred at a nitration unit during a chemical production plant explosion in Jilin, China, and about 100 tons of benzene and nitrobenzene leaked into the Songhua River [6]. Explosives are reactive substances that, upon initiation, undergo a rapid chemical transformation which produces gas, heat, and pressure, leading to an explosion. Explosive detection methods have been designed for various applications, including humanitarian demining, addressing environmental concerns like groundwater and soil remediation (as explosive residues pose a risk to human health), security screening, intelligence operations, criminal forensics, and rapid sample analysis, or the quantification of nitroaromatic compounds [7]. Another source of pollution is cars and exhaust fumes. Unfortunately, their widespread use has led to the contamination of soil and groundwater in the environment.

Chromatographic methods are often used to determine and separate nitro compounds in environmental matrices (most often water and soil samples). Mainly gas chromatography methods coupled with mass spectrometry (GC-MS) or an electron capture detector (GC-ECD) are suitable for the analysis of non-polar and volatile nitro compounds. Due to the incompatibility of most gas chromatography columns with water samples and due to the need to concentrate the analytes, it is necessary to choose a suitable extraction technique for extracting the analytes from the matrix.

To treat the water sample from environmental sources, conventional techniques such as Solid Phase Extraction (SPE) [8,9], Dispersive Solid Phase Extraction (dSPE) [10], Solid Phase Microextraction (SPME) [11–14], Switchable Hydrophilicity Solvent-Based Microextraction (SHS-ME) [15], Microextraction by Packed Sorbent (MEPS) [16,17], Stir Bar Sorptive Extraction (SBSE) [18], Dispersive Liquid–Liquid Microextraction (DLLME) [19–21], Single-drop Microextraction (SDME) [22–24], and Hollow Fiber Solid–Liquid Phase Microextraction (HF-SLPME) [12,24,25] were used. SDME is an attractive extraction technique mainly because of its low cost and ease of use. Finally, it follows the principles of green analytical chemistry, as only a few microliters of the extraction solvent are consumed for one extraction. The most commonly used SDME techniques include Headspace Single-drop Microextraction (HS-SDME), in which the extraction solvent does not come into direct contact with the sample but is suspended on the tip of a microsyringe in the "head-space" phase above the liquid sample surface [26–29], and Direct Immersion Single-drop Microextraction (DI-SDME), where the microdrop of the extraction solvent is directly immersed in the liquid sample [20–32]. As an extraction solvent, toluene [23,31], chlorobenzene [30], butyl acetate [26], n-octanol [32], N,N-dimethylformamide [28], or their combination, specifically toluene and butyl acetate [33] and benzyl alcohol and ethyl acetate [24] were already reported. A modern trend in SDME techniques is to replace traditional organic solvents and chlorinated solvents, such as chlorobenzene or chloroform, with greener alternatives with less impact on the environment [22,27,29]. Ecological solvents for microextraction techniques are a relatively new topic. New and innovative applications of deep eutectic solvents (DESs), switchable hydrophilicity solvents (SHSs), and supramolecular solvents (SUPRASs) in sample preparation are gaining increasing attention [34].

The main aim of the presented study is to develop a green analytical method based on SDME connected with GC and ECD detection for the simultaneous determination of nitro compounds from various types of environmental water and forensic rinse water samples. In this paper, the authors aim to present the optimized extraction method for different types of water samples and to optimize conditions for the variability of matrices. To our knowledge, detailed optimizing microextraction conditions and study of matrix effects for the analysis of nitro compounds for forensic rinse water samples have not been published so far.

2. Results and Discussion

For liquid phase samples, such as various types of environmental water and forensic rinse water, direct liquid extraction of nitro compounds is a promising sample pretreatment technique. Nowadays, there is a great emphasis on the greenness of the methods used; therefore, the minimization of extraction reagent leading to liquid microextraction techniques is a suitable option for testing the possibilities of analysis performed in a greener manner. The samples potentially contaminated with eight nitro compounds were analyzed, and the DI-SDME was used for extraction. Using the SDME technique, several extraction conditions have an important impact on the analytical method and the validation parameters, such as the choice of solvent, its volume, extraction time, the choice of mixing conditions, sample pH, and ionic strength.

2.1. Selection of Extraction Conditions

2.1.1. Selection of the Extraction Solvent and Its Volume

Working with aqueous solutions as samples, it is essential that the extractive solvent is immiscible with water so that it can form a microdrop in water. Several solvents were tested. Hexane as the extraction solvent is immiscible with water, but it was not able to form a stable microdrop in water. Of the potentially green solvents, n-butyl acetate was tested, as it is immiscible with water. This solvent created a stable microdroplet on the tip of the microsyringe. Unfortunately, during the extraction process, the volume of the microdroplet decreased, and after the end of the extraction, only a small fraction of the original volume of the droplet could be obtained as the extract. A volume of 0.2 μL was obtained from the original 1 μL n-butyl acetate drop; therefore, the solvent was not further applied. The next solvent tested was toluene. Toluene is immiscible with water and was able to form a stable microdroplet that maintained its original volume throughout the extraction time. The use of binary extraction solvents, which are mixtures of two extraction solvents, affects the polarity, density, and solubility of the final extraction product and thus affects the efficiency of the entire extraction procedure. In the next phase of testing, binary solvents were tested. Hexane and n-butyl acetate, respectively, were mixed with toluene in different proportions to eliminate the shortcomings of the listed solvents used individually. In the case of mixing hexane and toluene, two liquid immiscible phases that were not useful for the extraction were obtained. Combining n-butyl acetate with toluene produced a single liquid phase; unfortunately, the binary solvent droplet composition showed significant droplet volume loss during the extraction. A binary microdrop composed of toluene and n-butyl acetate in a ratio of 1:1 with a volume of 3 μL lost approximately one-third of its original volume during the extraction process. Therefore, only toluene was used as the extraction solvent in further experiments.

The droplet surface, through which the transfer of analytes from the sample to the extraction solvent occurs, is directly influenced by the quality of the solvent. The highest surface of the droplet will be increasing with the increase in the volume of the solvent. The calculated values of the vapor volume of the extraction solvent and the ability of the droplet to stay on the tip of the microsyringe throughout the extraction time were studied. For a 3 μL microdrop of toluene, under the selected pressure and temperature conditions of the subsequent GC analysis, the vapor volume was 442 μL for the given split/splitless inlet arrangement with the single-tapered liner type. The volume of the single tapered liner is 900 μL, so filling it with the injected extract to about half is suitable for injection to the chromatographic system. A drop with a volume greater than 3 μL did not remain on the microsyringe during the entire extraction time and would not be suitable for injection at the given inlet setting and temperature program; therefore, extractions were tested with drops at volumes of 1, 2, and 3 μL, while the extraction time remained constant at 10 min. Figure 1A shows the dependence of analyte recoveries on the droplet volume at the fixed extraction time of 10 min. For all analytes, by increasing the droplet volume, higher recovery values were obtained. Figure 1 also reveals the fundamental problem of SDME under the given conditions, which is repeatability. It was necessary to search conditions through optimization so that the system will be repeatable and robust.

2.1.2. Selection of the Time of the Extraction

The extraction time is the time during which the analytes are extracted into the extraction solvent. Too short an extraction time can lead to an incomplete transfer of analytes and thus significantly reduce the efficiency of the method. Conversely, too long an extraction time can lead to the deformation of the drop or its partial dissolution in the water sample. Therefore, it was necessary to choose an appropriate extraction time between these two extreme situations. The optimal time of the extraction was tested between 10 and 40 min. The highest peak areas of tested analytes were obtained using the extraction time of 35 min. Therefore, it was selected for further studies. The lower recoveries (also peak areas of analytes) in the 40 min extraction were caused by the accumulation of air bubbles in the droplet, which caused a part of it to break off and did not allow it to recover its entire volume. The intensity of mixing and the time the drop spent in the water could have contributed to the formation of tiny bubbles. All analytes showed a trend of increasing recoveries with increasing extraction time up to 35 min. Isolated drops could be caused by the break-off of a small volume of the drop or the formation of bubbles. Figure 1B shows the dependence of the recovery of tested analytes on the time of the extraction.

2.1.3. Selection of Stirring Intensity

The solution must be stirred throughout the extraction process to achieve a better transfer of analytes into the extraction microdrop. The optimization of this step is extremely important, because it is a micro-scale extraction, and the volume of the solvent is small, and it is placed in the fixed position in the sample throughout the extraction. It was expected that with an increase in the intensity of stirring, the peak areas of the analytes would also increase. The intensity of stirring was tested at the stirring rates between 0 and 400 rpm. At no stirring conditions (0 rpm), minimal recoveries (as well as peak areas of the analytes) were recorded, and at stirring intensity above 400 rpm, the droplet did not stay on the microsyringe tip for the entire extraction time and fell from the microsyringe tip. The highest recoveries (peak areas) were achieved at 300 rpm; the drop was stable during the entire extraction, and it was possible to load back its entire volume. By increasing the stirring intensity, a better mixing of the solution was achieved, but it led to the accelerated formation of bubbles in the drop, and thus, it was not always possible to load the entire volume of the drop back into the microsyringe. The repeatability of the experiments at the stirring intensity of 100 rpm was not acceptable due to high RSDs, which is mainly due to the instability of the drop at the needle tip. Figure 1C shows the dependence of the recovery of tested analytes on the stirring intensity. Further experiments were performed at 300 rpm.

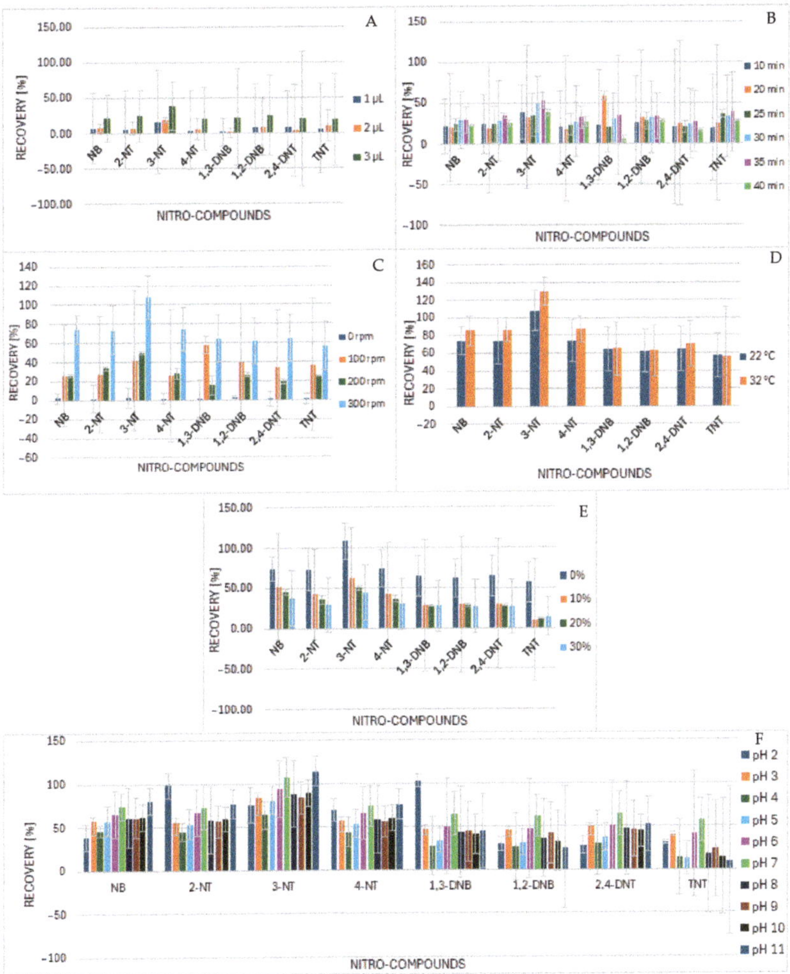

Figure 1. Dependence of the recovery of individual nitro compounds on the (**A**) volume of the microdrop, (**B**) the time of the extraction, (**C**) the stirring intensity of the extraction process, (**D**) the extraction temperature, (**E**) the addition of NaCl, and (**F**) the change in pH.

2.1.4. Selection of the Extraction Temperature

Increasing the temperature leads to greater solubility of the analytes in the sample but also to a decrease in the viscosity of the extraction solvent. In DI-SDME, this contributes to a higher instability of the drop, which causes a partial loss of volume due to its mixing with the aqueous phase or its complete detachment. The extraction temperature was examined at 22 °C, 32 °C, and 42 °C ± 1 °C. It was assumed that the transfer of analytes into the extraction solvent would improve with increasing temperature, which was confirmed through experiments. At a temperature of 32 °C, slightly higher analyte recoveries were achieved in comparison to experiments at the laboratory temperature. However, at an elevated temperature of 32 °C, it was not possible to regularly and reliably load back the entire volume of the drop into the microsyringe, and there was also an increased formation of tiny bubbles in the drop, which led to a worsening of the repeatability as well as success of the extraction. On average, only one out of four extractions were successful. In all other cases, the microdrop fell down from the microsyringe needle tip. At a temperature of 42 °C, a microdrop of the extraction solvent was not able to stay on the tip of the microsyringe during the entire extraction time. This phenomenon can be attributed to the fact that with increasing temperature, the viscosity of toluene decreases, and its mixing with the aqueous phase occurs, as well as the increased formation of bubbles in the drop. Satisfactory RSD values (15–26%) were achieved at 22 °C, which we attribute to less bubble formation in the drop and in the water sample, as well as a higher viscosity of the extraction solvent, which led to an increase in its stability. The highest extraction recoveries (peak areas of studied analytes) were achieved at 32 °C. The extraction temperature we continued to work with was 22 °C, although with lower analyte peak areas of the analytes, which resulted in a much higher extraction success rate of performing the extraction without a drop falling off the microsyringe needle tip; thus, the repeatability and robustness of the method was the highest (Figure 1D).

2.1.5. Addition of Salt

The addition of salt to the sample increases the ionic strength, which affects the solubility of organic analytes. The salt effect occurs, causing water molecules to interact with salt ions. Theoretically, this interaction disrupts the hydrogen bridges formed by the analytes, thereby increasing the transition of the analytes from the aqueous phase to the organic phase. The extractions were performed without the addition of salt and then with the addition of NaCl at concentrations of 10%, 20%, and 30% (w/v). The highest extraction recoveries were achieved without the addition of salt. Figure 1E shows the effect of NaCl addition on the recovery of individual nitro compounds. The graph shows that the highest recoveries were achieved in extractions without the addition of salt, while the lowest recoveries were achieved with the increase in the addition of salt. Also, less droplet stability after salt additions was observed, leading to reduced extraction success. No addition of NaCl was selected for the DI-SDME in the next studies.

2.1.6. Selection of the Working pH

Waters of different origins have different pH values. The deionized water that was used during the extraction had a pH value of 6. In contrast, sea water had a pH value of approximately 8, while mineral water and bottled water intended for consumption have pH values in the range of 5 to 10, depending on the manufacturer and the place of origin of the water. When determining the pH range in which the extractions will be performed, it was assumed that when analyzing real post-explosion products, the pH of the sample may be more acidic or basic than that of environmental water. Therefore, the study with water samples within the values of pH from 3 to 11 was performed. Aqueous solutions of citric acid and KOH were used to adjust the pH. The highest recoveries of analytes were achieved in a water sample with a pH value of 6, i.e., in deionized water without pH adjustment. In principle, significant differences between the extractions performed either in an acidic or alkaline environment were not noticed. The lowest RSD values were recorded for deionized water; however, in an alkaline environment of pH 8 to 11, an increase in the RSD values of extractions compared to the RSD values of extractions in an acidic environment was observed. In conclusion, the precision of the extraction is better in neutral and acidic environments. Therefore, it is necessary to evaluate the results of quantitative analysis for samples outside the optimal pH range by performing method validation for other pHs by using matrix-matched standards (Figure 1F).

2.2. *Validation Process*

The validation of the DI-SDME GC-ECD method was established in terms of linearity, the dynamical liner range, precision, accuracy, the limit of detection (LOD), and the limit of quantification (LOQ). To calculate the validation parameters, three calibration curves in matrices of deionized water, tap water, and seawater were performed. The capacity of the detector cell was no longer sufficient for higher concentrations. For analytes in deionized water, LOD values dependent on the individual nitro compounds ranged from 0.01 to 0.09 μg/L and LOQ values from 0.03 to 0.31 μg/L. The coefficients of determination were in the range of 0.9961 to 0.9994 for all analytes. Analytes in the tap water matrix had LOD values from 0.01 to 0.06 μg/L and LOQ values from 0.03 to 0.19 μg/L. The coefficients of determination were in the range of 0.9977 to 0.9993. For analytes in the seawater matrix, LOD values ranged from 0.01 to 0.03 μg/L and LOQ values from 0.03 to 0.11 μg/L. The coefficients of determination had a value of 0.9974 to 0.9992. For analytes in the model forensic rinse water matrix, the LOD values range from 0.03 to 0.11 μg/L and LOQ values from 0.11 to 0.38 μg/L. The coefficients of determination had a value of 0.9611 to 0.9965. Tables 1 and 2 shows the validation parameters for different types of water samples.

Table 1. Linearity, LOD, and LOQ for deionized, tap, sea, and forensic rinse water matrices.

Analyte	Deionized Water Matrix			Tap Water Matrix			Seawater Matrix			Forensic Rinse Water Matrix		
	Linearity R^2	LOD (μg/L)	LOQ (μg/L)	Linearity R^2	LOD (μg/L)	LOQ (μg/L)	Linearity R^2	LOD (μg/L)	LOQ (μg/L)	Linearity R^2	LOD (μg/L)	LOQ (μg/L)
NB	0.9973	0.02	0.08	0.9989	0.01	0.04	0.9991	0.01	0.02	0.9926	0.05	0.16
2-NT	0.9984	0.05	0.15	0.9993	0.06	0.19	0.9992	0.02	0.08	0.9965	0.03	0.11
3-NT	0.9961	0.08	0.28	0.9986	0.06	0.19	0.9992	0.01	0.03	0.9965	0.05	0.15
4-NT	0.9973	0.09	0.31	0.9974	0.02	0.07	0.9990	0.02	0.07	0.9611	0.11	0.38
1,3-DNB	0.9992	0.03	0.10	0.9986	0.01	0.05	0.9985	0.03	0.10	0.9873	0.05	0.17
1,2-DNB	0.9992	0.01	0.03	0.9946	0.01	0.03	0.9975	0.01	0.03	0.9820	0.06	0.21
2,4-DNT	0.9982	0.02	0.07	0.9975	0.02	0.06	0.9985	0.03	0.11	0.9848	0.06	0.19
TNT	0.9994	0.05	0.18	0.9977	0.01	0.05	0.9974	0.01	0.03	0.9620	0.09	0.31

Notes: R^2—determination coefficient, LOD—limit of detection, and LOQ—limit of determination. Analyte name abbreviations: NB—Nitrobenzene, 2-NT—2-Nitrotoluene, 3-NT—3-Nitrotoluene, 4-NT—4-Nitrotoluene, 1,3-DNB—1,3-Dinitrobenzene, 1,2-DNB—1,2-Dinitrobenzene, 2,4-DNT—2,4-Dinitrotoluene, and TNT—Trinitrotoluene.

2.3. *Study of the Matrix Effect*

The matrix effect may manifest either an increase or decrease in the detector signal compared to the response from solvent solutions of the analytes. The matrix factor (MF) was calculated for each studied nitro compound by comparing the peak area of the analyte in the matrix-matched solution vs. the peak area of the nitro compound in distilled water. The MF was investigated in tap water, in seawater, and in model forensic rinse water. A positive value of the matrix effect corresponds to an amplification of the matrix-induced signal, while a negative value corresponds to a signal attenuation effect. In general, matrix effects can be classified as weak (0–20%), moderate (20–50%), and strong (>50%). The matrix factors are summarized in Table 3 and Figure 2.

Table 2. Recovery and relative standard deviation for deionized, tap, sea, and forensic rinse water matrices.

Analyte	Deionized Water Matrix		Tap Water Matrix		Seawater Matrix		Forensic Rinse Water Matrix	
	Recovery (%)	RSD (%)	Recovery (%)	RSD (%)	Recovery (%)	RSD (%)	Recovery (%)	RSD (%)
NB	74.14	15.59	90.48	10.31	71.11	14.61	69.04	15.78
2-NT	73.47	26.03	81.78	11.86	63.94	23.59	35.65	51.06
3-NT	107.93	22.73	115.20	24.64	99.88	20.34	30.51	30.06
4-NT	73.90	23.16	91.89	28.53	62.15	24.81	36.98	4.20
1,3-DNB	64.51	24.80	101.11	17.61	61.99	13.52	52.60	2.17
1,2-DNB	62.16	23.90	101.63	20.78	63.49	17.87	57.79	14.68
2,4-DNT	64.63	25.09	106.26	19.54	63.56	9.86	63.39	21.24
TNT	56.46	24.87	100.12	20.25	60.01	14.31	101.61	48.56

Notes: RSD—relative standard deviation. Analyte name abbreviations: NB—Nitrobenzene, 2-NT—2-Nitrotoluene, 3-NT—3-Nitrotoluene, 4-NT—4-Nitrotoluene, 1,3-DNB—1,3-Dinitrobenzene, 1,2-DNB—1,2-Dinitrobenzene, 2,4-DNT—2,4-Dinitrotoluene, and TNT—Trinitrotoluene.

Table 3. Matrix factors of individual analytes in different types of water samples.

Analyte	MF in Tap Water (%)	MF in Seawater (%)	MF in Forensic Rinse Water (%)
NB	−31.13	18.64	−28.51
2-NT	−25.83	33.40	−23.84
3-NT	−31.69	18.61	−36.30
4-NT	−24.40	36.24	−55.52
1,3-DNB	−52.15	11.82	−2.16
1,2-DNB	−50.67	5.62	−0.89
2,4-DNT	−50.66	7.56	−0.93
TNT	−39.27	2.58	−0.91

Notes: MF—matrix factor. Analyte name abbreviations: NB—Nitrobenzene, 2-NT—2-Nitrotoluene, 3-NT—3-Nitrotoluene, 4-NT—4-Nitrotoluene, 1,3-DNB—1,3-Dinitrobenzene, 1,2-DNB—1,2-Dinitrobenzene, 2,4-DNT—2,4-Dinitrotoluene, and TNT—Trinitrotoluene.

As shown in Figure 2, moderate and high matrix effects were detected for analytes in tap water. Weak matrix effects were detected for most of the analytes in the seawater matrix, with the exception of 2-NT and 4-NT, where the matrix effect was moderate. Moderate matrix effects were found for 3-NT and 4-NT in the model forensic rinse water matrix, and weak matrix effects were found for the other analytes. We can evaluate that for nitrotoluene isomers in all matrices, the matrix effects were significantly lower than those for the other analytes.

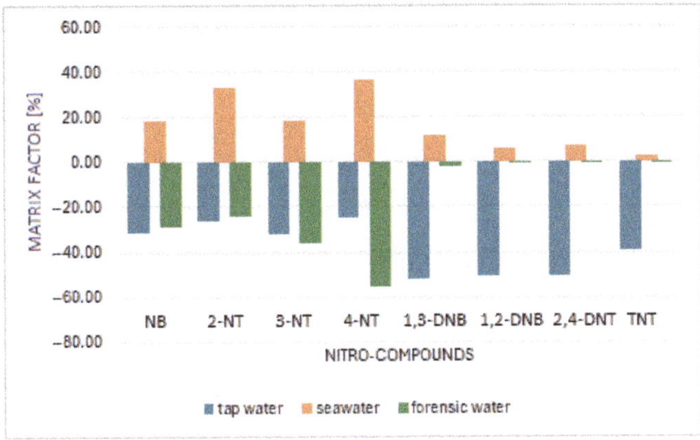

Figure 2. Matrix factors of individual analytes in different types of water samples.

2.4. Real Sample Analysis

The applicability of the method was tested via real sample analyses under optimized conditions in various types of water samples: forensic, seawater, and tap water. Four samples of tap water were collected from different points of Slovakia, and one seawater sample and one forensic rinse water sample were analyzed. Figure 3 shows the result of the chromatographic analyses. No levels above the limit of detection were found.

2.5. Greenness of the DI-SDME Method

The evaluation of the greenness of the method was conducted according to AGREE, an analytical greenness calculator software (https://agree-index.anvil.app/ accessed on 26 March 2025); AGREEprep, an analytical greenness metric for sample preparation (https://agreeprep.anvil.app/ accessed on 26 March 2025); and the analytical eco-scale assessment (AESA) method [35]. The AGREE software consists of twelve questions; AGREEprep is based on 10 categories of green analytical chemistry, and according to the answers, it rates the greenness of the method from 0 to 100% [36,37]. The AESA method deducts points for the use of toxic chemicals, energy consumption, and the amount of waste produced, among others [38]. Figure 4A shows the results of the method evaluation with AGREE.

The resulting evaluation of the greenness of the method according to AGREE was 70%. The main shortcomings identified were off-line sampling and analysis (1, 3), manual injection (5), and GC-ECD as a separation technique. The sample volume was suboptimal for greenness and noted that the method lacked automation. If we consider the evaluation of the parameters related exclusively to the extraction method, we can evaluate that the technique is green, and even the use of toluene in a small volume of 3 µL did not detract from the evaluation. Figure 4 shows the results of the method evaluation with AGREE.

The assessment result according to AGREEprep is 71%. The limitations of the proposed method are as follows: the sample preparation is performed in the laboratory after sample collection and transportation, which affects the overall speed of sample preparation, and the use of the extraction solvent. Figure 4B shows the results of the method evaluation with AGREEprep.

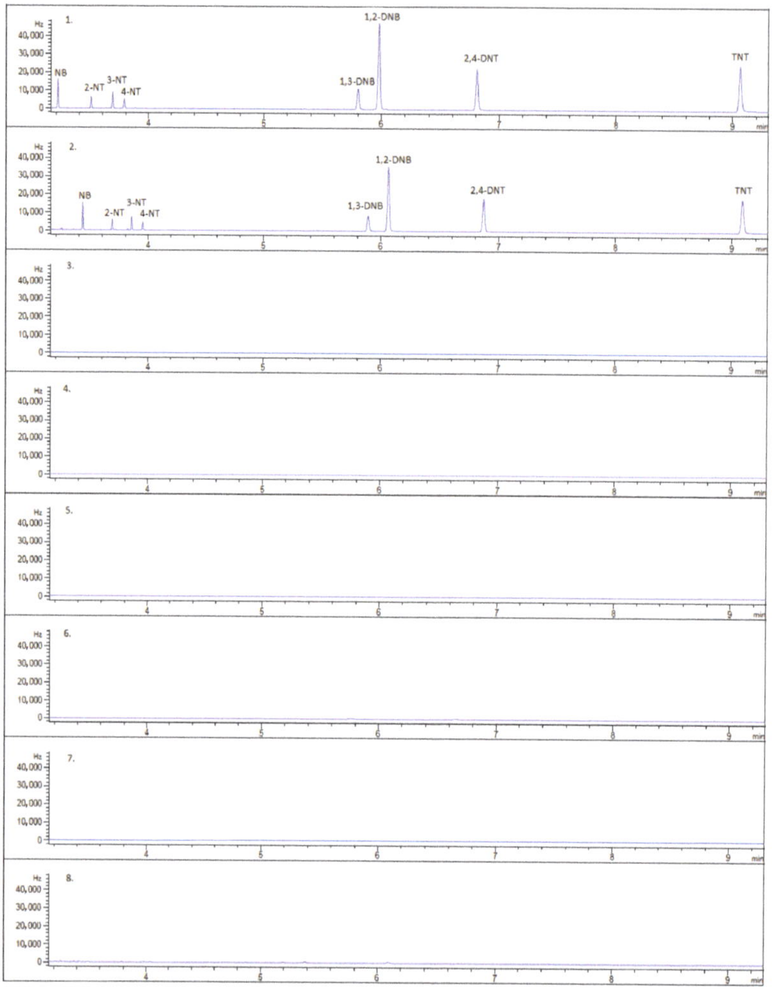

Figure 3. Chromatograms of real samples: (**1**) nitro compounds in the solvent (toluene), (**2**) the spiked water sample with nitro compounds, (**3**) the forensic rinse water sample, (**4**) the seawater sample, (**5**) the drinking water sample, (**6**) the drinking water sample, (**7**) the drinking water sample, (**8**) the drinking water sample.

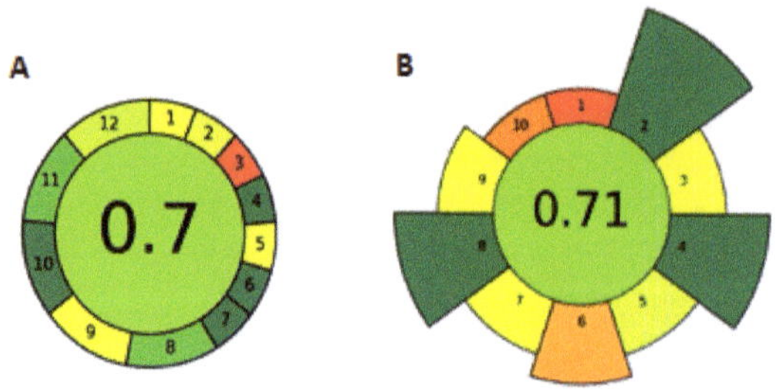

Figure 4. Evaluation of the greenness of the method according to (**A**) AGREE and (**B**) AGREEprep.

According to the AESA evaluation, the developed method reached 90 points. According to the authors of AESA [35], the number of points above 75 is rated as an excellent green method. The biggest shortcomings in the evaluations were the work with toluene and the danger associated with it, as well as the large volume of the analyzed sample. Another focus of the extraction technique on even greater greenness can be the automation of the technique and the use of greener solvents than toluene. Table 4 shows the results of the method evaluation according to AESA.

Table 4. Evaluation of the greenness of the method according to AESA.

Procedure Parameters	Hazard	Penalty Points
Reagents	Toluene (3 µL)	6
Energy	GC-ECD	1
	Magnetic stirrer	0
Occupational hazard	Toluene vapors	3
Waste	Water sample (0 mL)	0
Sum	100 − 10 =	90

2.6. Comparison of the Developed Method

The results obtained using the developed SDME-GC-µECD method were consistent with those from other sample preparation techniques, including MEPS, SPME, directly suspended solidified floating organic droplet microextraction (DS-SFOD), and SDME, and closely aligned with previously published data [13,16,22,39]. The developed method shows LOD and LOQ comparable or lower to those of existing approaches. Notably, it employs a minimal amount of sample and extraction solvent volume and multi-element analysis. These key advantages highlight the suitability for the determination of explosives. A comparative analysis between the proposed method and other relevant analytical techniques for explosive enrichment is presented in Table 5.

Table 5. Comparison of the developed method with methods reported earlier.

Sample Matrix	Analytes	Sample Preparation	Extraction Parameters	Instrumental Method	Recovery	LOD (LOQ)	Real Finds	References
Distilled and well water	2-CA, 2,5-DCA, 2-NT, 3-NT, 4-CNB, 2,5-DCNB, 3,4-DCNB	SPME	Sample size: 1.5 mL Time of extraction: 45 min Desorption time: 3 min Desorption temperature: 250 °C	GC-FID	73–119%	1–10 µg/L 30–50 µg/L	2-CA 0.06 mg/L 0.63 mg/L 2,5-DCA 0.36 mg/L 3,4-DCNB 0.08 mg/L 9.42 mg/L 2-NT, 3-NT, 4-CNB, 2,5-DCNB Not detected	[13]
Underground water	NB, TNT, Tetryl, 1,3,5-TNB, 4-ADNT, 1,3-DNB, 2,4-DNT, 2,6-DNT, 2-NT, 3-NT, 4-NT	MEPS	Sample size: 10 × 50 µL Washing solvent: 50 µL of water Elution solvent: 30 µL of MeOH	GC-MS	77.5–99.2%	0.014–0.828 pg/mL 0.046–2.732 ng/mL	NB 1.03 ng/mL 2-NT 0.38 ng/mL 3-NT 0.81 ng/mL 2,4-DNT 0.22 ng/mL	[16]

Table 5. *Cont.*

Sample Matrix	Analytes	Sample Preparation	Extraction Parameters	Instrumental Method	Recovery	LOD (LOQ)	Real Finds	References
Surface, tap, and well water	TNT, 2,4-DNT, 2,6-DNT	DS-SFOD	Sample size: 10 mL Extraction solvent: 40 µL of DES (Decanoic acid–borneol) Time of extraction: 30 min	HPLC-UV	89–102%	0.14–0.19 µg/L 0.5–0.6 µg/L	<LOD	[22]
Well and underground water	HMX, RDX, PETN, TNB, CL-20, Tetryl, TNT, 2,6-DNT, 3-NT, 2,4-DNT, 2-NT	d-SDME	Sample size: 10 mL Extraction temperature: 30 °C Extraction solvent: 50 µL (ferrofluid) Time of extraction: 30 min Dissolving solution: 500 µL of Acetonitrile	HPLC-UV	88–103.7%	0.22–0.91 µg/L 0.73–3 µg/L	RDX 2.3 µg/L TNB 0.7 µg/L 6.4 µg/L TNT 2.5 µg/L 12.4 µg/L 2,6-DNT 0.9 µg/L 9.8 µg/L	[39]
Deionized, tap, sea, and forensic rinse water	NB, 2-NT, 3-NT, 4-NT, 1.3-DNB, 1.2-DNB, 2.4-DNT, TNT	SDME	Sample size: 2 mL Extraction temperature: 22 °C Extraction solvent: 3 µL Toluene Time of extraction: 35 min	GC-µECD	57–115%	0.01–0.11 µg/L 0.03–0.38 µg/L	<LOD	This method

Notes: LOD—limit of detection, LOQ—limit of determination, DS-SFOD—directly suspended solidified floating organic droplet microextraction, SPME—solid phase microextraction, MEPS—microextraction by packed sorbent, d-SDME—directly suspended droplet microextraction, and SDME—single-drop microextraction. Analyte name abbreviations: 2-CA—2-Chloroaniline, 2,5-DCA—2,5-Dichloroaniline, 2-NT—2-Nitrotoluene, 3-NT—3-Nitrotoluene, 4-CNB—4-Chloronitrobenzene, 2,5-DCNB—2,5-Dichloronitrobenzene, 3,4-DCNB—3,4-Dichloronitrobenzene, NB—Nitrobenzene, TNT—2,4,6-Trinitrotoluene, Tetryl—2,4,6-Trinitrophenyl-Nmethylnitramine, 1,3,5-TNB—1,3,5-Trinitrobenzene, 4-ADNT—4-Amino-2,6-dinitrotoluene, 1,3-DNB—1,3-Dinitrobenzene, 2,4-DNT—2,4-Dinitrotoluene, 2,6-DNT—2.6-Dinitrotoluene, 4-NT—4-Nitrotoluene, HMX—Cyclotetramethylene tetranitramine, RDX—1,3,5-Trinitroperhydro-1,3,5-triazine, PETN—Pentaerythritol tetranitrate, TNB—1,3,5-Trinitrobenzene, CL-20—Hexanitrohexaazaisowurtzitane, 1,3-DNB—1,3-Dinitrobenzene, and 1,2-DNB—1,2-Dinitrobenzene.

The table also lists actual findings by other authors. Actual findings fall within the concentration range presented in this paper.

3. Experimental

3.1. Chemicals and Reagents

All individual standards of nitro compounds were purchased from different sources (Dr. Ehrenstorfer GmbH, Augsburg, Germany) and provided by the Forensic Institute, Bratislava, Slovakia with purity higher than 95.66%. Stock standard solutions of the individual nitro compound, at a concentration of 1 mg/mL, were prepared by weighing the nitro compound and dissolving it in 10 mL of methanol (VWR Chemicals, Missouri City, MO, USA). The working standard solution of nitro compounds was prepared in methanol by dilution and stored at 4 °C. The working solution contained 9 nitro compounds at a concentration of 0.1 ng/µL. The calibration standards were prepared via additional dilution with toluene (Merck KGaA, Darmstadt, Germany). Solvents with different polarities, such as acetone, chlorobenzene, chloroform (all from Centralchem, Bratislava, Slovakia), dichloromethane, n-pentane, n-hexane and ethyl-acetate (all from Merck KGaA, Darmstadt,

Germany), and acetonitrile (VWR Chemicals, Missouri City, MO, USA), were used as dilution solvents and for the preparation of working solutions. To obtain the optimum pH, citric acid (Mikrochem, Pezinok, Slovakia) and potassium hydroxide (Chemapol, Bratislava, Slovakia) were used. Sodium chloride (Centralchem, Bratislava, Slovakia) was added to optimize the addition of salt.

3.2. Samples and Procedures

The real tap water sample was collected in Bratislava, and seawater samples were taken from two sampling points on the island of Pag, Adriatic Sea (Croatia); the model forensic rinse water sample was prepared by swabbing the surface of the laboratory equipment with water moistened cotton swabs, thus imitating the real forensic sample collection at the site after an explosion. Real water samples were transferred to the analytical laboratory and stored in 1 L amber glass bottles in the fridge at 4 °C until DI-SDME-GC-ECD analysis.

The technique of DI-SDME was used to extract the analytes from the aqueous samples into the extract. A 10 μL microsyringe containing 3 μL of the extraction solvent was inserted into a 2 mL water sample contained in a 4 mL vial and subjected to magnetic stirring. The extraction solvent was ejected from the tip of the microsyringe, forming a drop on the microsyringe tip, and sample mixing was started. After the extraction time, the mixing machine was turned off, and the microdroplet was drawn back into the syringe and was injected into the GC system.

3.3. Instrumental Analysis and Data Evaluation

Chromatographic analyses were carried out on a gas chromatograph from Agilent Technologies, 6890N (Agilent Technologies, Little Falls, DE, USA). The chromatograph was equipped with a μECD (electron capture detector) and a split/splitless injector. Analytes were separated on a narrow-bore CP Sil 8 CB column with a stationary phase containing a 5% phenyl-polydimethylsiloxane polymer (15 m × i.d. 150 μm × 0.15 μm film thickness). SDMEs were performed on a magnetic stirrer LLG-uniSTIRRER 7. Extracts were injected in splitless mode using a 10 μL microsyringe. Hydrogen (4.0 Linde Technoplyn; Bratislava, Slovakia) was used as carrier gas at a constant flow rate of 1 mL/min. The injector temperature was set at 200 °C and used in splitless mode. The separations were carried out under temperature-programmed conditions. The initial temperature of the chromatographic oven was set at 90 °C and held for 1 min; subsequently, the temperature was increased to 130 °C at a 15 °C/min rate and held for 1 min. After that, the temperature was increased to 160 °C at a rate of 7 °C/min and held for 1 min, followed by a rate of 60 °C/min until 250 °C and held for 1 min. The analysis run time was 12.45 min. The detector temperature was set at 250 °C; nitrogen (5.0 Linde Technoplyn; Bratislava, Slovakia) was used as the makeup gas at a constant flow of 60 mL/min. The pH values of the solutions were measured on a Cyberscan pH 510 pH meter (Eutech Instruments, Vernon Hills, IL, USA). Sartorius Analytic MC1 scales (Sartorius, Göttingen, Germany) were used for weighing standards.

3.4. Method Validation

Deionized water, tap water, seawater, and model forensic rinse water samples were used as blanks for the preparation of matrix-matched standard solutions in the recovery study. The spiked standard solutions were prepared by adding working solutions of the respective nitro compounds using the optimized DI-SDME method.

3.4.1. Repeatability, Linearity, Limit of Detection, and Limit of Quantification

RSD values were used to evaluate the repeatability of the method. Each extraction was repeated three times. The errors were calculated based on the standard deviation. The linearity of the method was evaluated using matrix-matched calibration curves through the spiking of blank sample extracts in the range from 0.5 to 250 μg/L. To evaluate the linearity, the coefficient of determination R^2 was calculated. The limit of detection (LOD) and the limit of quantification (LOQ) were calculated from the slope of the calibration curve.

$$\text{LOD} = (3.3 * \text{standard deviation of the blank/slope of the calibration curve}) \quad (1)$$

$$\text{LOQ} = 3.3 * \text{LOD} \quad (2)$$

3.4.2. Matrix Effects

The matrix effect was studied by calculating the matrix factor (MF) for each nitro compound. The MFs were calculated by comparing the peak area of the analyte in the matrix-matched solution vs. the peak area of the nitro compound in distilled water. The matrix-matched solution was prepared by using 3 μL of the matrix extract and 1 μL of the standard solution in toluene of known concentration.

$$\text{MF} = (\text{peak area in matrix-matched solution/peak area in extract from distilled water} - 1) * 100 \quad (3)$$

4. Conclusions

This paper presents a broad study on the development of a green, environmentally friendly SDME method, with minimal consumption of solvents, energy, and waste generation. Eight nitro compounds were monitored in the water samples. Analyses were carried out using GC-μECD, and samples of deionized water and real water samples were analyzed. As part of method validation, linearity, precision, the limit of detection, and the limit of quantification were evaluated. The developed DI-SDME method can be reliably used for the analysis of selected analytes in water samples of monitored matrices. Matrix effects were evaluated, demonstrating that the matrix had an impact on the calibration process, and matrix-matched calibration is recommended. The greenness of the developed method was also evaluated with these techniques: AGREE (analytical greenness calculator software), AGREEprep, and the AESA method. With AGREE, the method reached 70%, while with AGREEprep, it reached 71%, and with AESA, the proposed method reached 90 points and was evaluated as an excellent green analysis.

Author Contributions: Conceptualization, S.H.; data curation, T.P.; formal analysis, T.P.; funding acquisition, T.P. and S.H.; investigation, S.O.; methodology, T.P., S.O. and S.H.; project administration, S.H.; resources, S.H.; supervision, S.H.; validation, T.P.; visualization, T.P.; writing—original draft, T.P.; writing—review and editing, S.H. All authors have read and agreed to the published version of the manuscript.

Funding: This work was supported by the Slovak Research and Development Agency under the Contract No. APVV-19-0149. This work was supported by the call for doctoral students and young researchers of Slovak University of Technology in Bratislava to start a research career (Grant 23-04-05-A).

Institutional Review Board Statement: Not applicable.

Informed Consent Statement: Not applicable.

Data Availability Statement: All relevant data are included in the paper.

Conflicts of Interest: The authors declare no conflicts of interest.

References

1. Ju, K.; Parales, R.E. Nitroaromatic compounds, from synthesis to biodegradation. *Microbiol. Mol. Biol. Rev.* **2010**, *74*, 250–272. [CrossRef] [PubMed]
2. Calza, P.; Massolino, C.; Pelizzetti, E.; Minero, C. Solar driven production of toxic halogenated and nitroaromatic compounds in natural seawater. *Sci. Total Environ.* **2008**, *398*, 196–202. [CrossRef] [PubMed]
3. Winkler, R.; Hertweck, C. Biosynthesis of nitro compounds. *Chembiochem* **2007**, *8*, 973–977. [CrossRef] [PubMed]
4. White, P.A.; Claxton, L.D. Mutagens in contaminated soil: A review. *Mutat. Res. Rev. Mutat. Res.* **2004**, *567*, 227–345. [CrossRef]
5. Zhao, Y.B.; Hayeck, N.; Saliba, N.A.; Schreiner, C.; Zennegg, M.; Jiang, F.; Figi, R.; Bleiner, D.; Wang, J. Any long-term effect on the Beirut port explosion on the airborne particulate matter? *Aerosol. Air Qual. Res.* **2023**, *23*, 220395. [CrossRef]
6. Lague, D. China blames oil company for benzene spill in river. *New York Times*, 25 November 2005.
7. Zhang, W.; Tang, T.; Shi, A.; Bao, L.; Shen, Y.; Shen, R.; Ye, Y. Recent Developments in spectroscopic techniques for the detection of explosives. *Materials* **2018**, *11*, 1364. [CrossRef]
8. Bednar, A.J.; Russell, A.L.; Hayes, C.A.; Jones, W.T.; Tackett, P.; Splichal, D.E.; Georgian, T.; Parker, L.V.; Kirgan, R.A.; Macmillan, D.K. Analysis of munitions constituents in groundwater using a field-portable GC–MS. *Chemosphere* **2015**, *87*, 894–901. [CrossRef]
9. Bünning, T.H.; Strehse, J.S.; Hollmann, A.C.; Bötticher, T.; Maser, E.A. Toolbox for the determination of nitroaromatic explosives in marine water, sediment, and biota samples on femtogram levels by GC-MS/MS. *Toxics* **2021**, *9*, 60. [CrossRef]
10. Zarei, A.R.; Nedaei, M.; Ghorbanian, S. Application of deep eutectic solvent based magnetic colloidal gel for dispersive solid phase extraction of ultra-trace amounts of some nitroaromatic compounds in water samples. *J. Mol. Liq.* **2017**, *246*, 58–65. [CrossRef]
11. Akbarian, M.; Gholamalizadeh, A.; Ahmar, H.; Banitaba, M.H. Application of poly 3,4-ethylenedioxythiophene and gold nanoparticles composite on a gold wire as a coating for determination of nitroaromatics in soil using cold fiber solid phase microextraction combined with gas chromatography. *Environ. Nanotechnol. Monit. Manag.* **2023**, *20*, 100804. [CrossRef]
12. Psillakis, E.; Mantzavinos, D.; Kalogerakis, N. Development of a hollow fibre liquid phase microextraction method to monitor the sonochemical degradation of explosives in water. *Anal. Chim. Acta* **2004**, *501*, 3–10. [CrossRef]
13. Suchana, S.; Passeport, E. Optimization of a solid-phase microextraction technique for chloro- and nitro- substituted aromatic compounds using design of experiments. *J. Chromatogr. A* **2020**, *1621*, 461083. [CrossRef] [PubMed]
14. Zang, X.; Zhang, X.; Chang, Q.; Li, S.; Wang, C.; Wang, Z. Metal–organic framework UiO-67-coated fiber for the solid-phase microextraction of nitrobenzene compounds from water. *J. Sep. Sci.* **2016**, *39*, 2770–2776. [CrossRef] [PubMed]
15. Rameshgar, J.; Hasheminasab, K.S.; Adlnasab, L.; Ahmar, H. Switchable-hydrophilicity solvent-based microextraction combined with gas chromatography for the determination of nitroaromatic compounds in water samples. *J. Sep. Sci.* **2017**, *40*, 3015–3182. [CrossRef]
16. Dhingra, G.; Bansal, P.; Dhingra, N.; Rani, S.; Malik, K.A. Development of a microextraction by packed sorbent with gas chromatography-mass spectrometry method for quantification of nitroexplosives in aqueous and fluidic biological samples. *J. Sep. Sci.* **2017**, *41*, 639–647. [CrossRef]
17. Hrouzková, S.; Pócsová, T.; Lelkesová, T.; Ulbrich, P. Determination of ethylene glycol dinitrate in environmental and forensic water samples using microextraction by Packed Sorbent followed by gas chromatography. *Separations* **2023**, *10*, 227. [CrossRef]
18. Galmiche, M.; Colin, A.; Clavos, M.C.; Pallez, C.; Rosin, C.; Dauchy, X. Determination of nitroaromatic explosive residues in water by stir bar sorptive extraction-gas chromatography-tandem mass spectrometry. *Anal. Bioanal. Chem.* **2021**, *413*, 159–169. [CrossRef]
19. Sobhi, H.R.; Kashtiaray, A.; Farahani, H.; Javaheric, M.; Ganjalic, M.R. Quantitation of mononitrotoluenes in aquatic environment using dispersive liquid–liquid microextraction followed by gas chromatography–flame ionization detection. *J. Hazard. Mater.* **2010**, *175*, 279–283. [CrossRef]
20. Sobhi, H.R.; Kashtiaray, A. Tandem use of solid-phase extraction and dispersive liquid-liquid microextraction for the determination of mononitrotoluenes in aquatic environment. *J. Sep. Sci.* **2011**, *34*, 1035–1040. [CrossRef]
21. Zhang, D.; Zeng, X.; Yu, Z.; Shenga, G.; Fua, J. Determination of nitrobenzenes and nitrochlorobenzenes in water samples using dispersive liquid-liquid microextraction and gas chromatography-mass spectrometry. *Anal. Methods* **2011**, *3*, 2254–2260. [CrossRef]
22. Moazzen, S.; Zarei, A.R.; Mardi, K. Green sample preparation based on directly suspended droplet microextraction using deep eutectic solvent for ultra-trace quantification of nitroaromatic explosives by high performance liquid chromatography. *J. Anal. Chem.* **2021**, *76*, 1296–1304. [CrossRef]
23. Psillakis, E.; Kalogerakis, N. Solid-phase microextraction versus single-drop microextraction for the analysis of nitroaromatic explosives in water samples. *J. Chromatogr. A* **2001**, *938*, 113–120. [CrossRef] [PubMed]
24. Sarafraz-Yazdi, A.; Es'haghi, Z. Comparison of hollow fiber and single-drop liquid-phase microextraction techniques for HPLC determination of aniline derivatives in water. *Chromatographia* **2006**, *63*, 563–569. [CrossRef]

25. Fayazi, M.; Ghanei-Motlagh, M.; Taherb, M.A. Combination of carbon nanotube reinforced hollow fiber members microextraction with gas chromatography-mass spectrometr25y for extraction and determination of some nitroaromatic explosives in environmental water. *Anal. Methods* **2013**, *6*, 1474–1480. [CrossRef]
26. Ma, Z.; Zhao, T.; Cui, S.; Zhao, X.; Fan, Y.; Song, J. Determination of ethyl carbamate in wine by matrix modification-assisted headspace single-drop microextraction and gas chromatography–mass spectrometry technique. *Food Chem.* **2022**, *373*, 131573. [CrossRef]
27. Mehrevar, A.; Feizbakhsh, A.; Sarafi, A.H.M.; Konoz, E.; Faraji, H. Deep eutectic solvent-based headspace single-drop microextraction of polycyclic aromatic hydrocarbons in aqueous samples. *J. Chromatogr. A* **2020**, *1632*, 461618. [CrossRef]
28. Qin, B.; Wang, X.; Tang, L.; Wang, S.; Shi, Y.; Zhao, L.; Jiang, H. Comparative study of headspace and headspace single drop microextraction combined with GC for the determination of methanol in wine. *J. Chromatogr. A* **2022**, *1673*, 463079. [CrossRef]
29. Triaux, Z.; Petitjean, H.; Marchioni, E.; Boltoeva, M.; Marcic, C. Deep eutectic solvent–based headspace single-drop microextraction for the quantification of terpenes in spices. *Anal. Bioanal. Chem.* **2020**, *412*, 933–948. [CrossRef]
30. Chen, Y.; Liu, H.; Peng, J.; Zhou, H.; Luan, T. Determination of polybrominated diphenyl ethers and metabolites by singledrop microextraction and GC–MS/MS. *SN Appl. Sci.* **2020**, *2*, 986. [CrossRef]
31. Mourabit, F.; M'hamed, A.; Houssien, A.; Mourabit, N. Single-drop microextraction application for the rapid and sensitive determination of endosulfan and its metabolites in water samples. *E3S Web Conf.* **2024**, *502*, 04005. [CrossRef]
32. Saraji, M.; Javadian, S. Single-drop microextraction combined with gas chromatography-electron capture detection for the determination of acrylamide in food samples. *Food Chem.* **2019**, *274*, 55–60. [CrossRef] [PubMed]
33. Chullasat, K.; Huang, Z.; Bunkoed, O.; Kanatharana, P.; Lee, H.K. Bubble-in-drop microextraction of carbamate pesticides followed by gas chromatography-mass spectrometric analysis. *Microchem. J.* **2020**, *155*, 104666. [CrossRef]
34. Carasek, E.; Bernardi, G.; Morelli, D.; Merib, J. Sustainable green solvents for microextraction techniques: Recent developments and applications. *J. Chromatogr. A* **2021**, *1640*, 461944. [CrossRef] [PubMed]
35. Gałuszka, A.; Migaszewskia, Z.; Konieczka, P.; Namieśnik, J. Analytical Eco-Scale for assessing the greenness of analytical procedures. *Trends Anal. Chem.* **2012**, *37*, 61–72. [CrossRef]
36. Pena-Pereira, F.; Wojnowski, W.; Tobiszewski, M. AGREE—Analytical GREEnness metric approach and software. *Anal. Chem.* **2020**, *92*, 10076–10082. [CrossRef]
37. Wojnowski, W.; Tobiszewski, M.; Pena-Pereira, F.; Psillakis, E. AGREEprep–Analytical greenness metric for sample preparation. *Trends Anal. Chem.* **2022**, *149*, 116553. [CrossRef]
38. Gałuszka, A.; Migaszewskia, Z.; Namieśnik, J. The 12 principles of green analytical chemistry and the SIGNIFICANCE mnemonic of green analytical practices. *Trends Anal. Chem.* **2013**, *50*, 78–84. [CrossRef]
39. Zarei, A.R.; Nedaei, M.; Ghorbanian, S.A. Ferrofluid of magnetic clay and menthol based deep eutectic solvent: Application in directly suspended droplet microextraction for enrichment of some emerging contaminant explosives in water and soil samples. *J. Chromatogr. A* **2018**, *1553*, 32–42. [CrossRef]

molecules

MDPI

Article

Metal Uptake by Birches and Scots Pines Grown on a Porcelain Landfill

Michaela Zeiner [1,*], Viktor Sjöberg [1] and Helena Olsman [2]

1 Man-Technology-Environment Research Centre, School of Science and Technology, Örebro University, Fakultetsgatan 1, 70182 Orebro, Sweden
2 Eurofins European Competence Centre for PFAS, Eurofins Food & Feed Testing Sweden AB, Sjöhagsgatan 3, 53140 Lidkoping, Sweden
* Correspondence: michaela.zeiner@oru.se

Abstract

Potentially toxic elements (PTEs) have steadily become a serious environmental problem, especially regarding brownfields chosen for reuse, e.g., as a residential area. "Norra Hamnstaden" in Lidköping (Sweden) has a long history of industrial activity, including porcelain production with the resultant industrial waste deposited close by resulting in elevated levels of metals used for porcelain glazes, especially lead. To estimate the bioavailability of twelve PTEs (As, Ba, Pb, Cd, Co, Cu, Cr, Mn, Mo, Ni, V, Zn), their uptake by birches (*Betula pendula*) as well as Scots pines (*Pinus sylvestris*) was investigated through analyzing their leaves. Sampling was carried out on five trees once per month in the period from May to August. Different uptake patterns were observed for birches and pines, for the latter even varying with age. The birch samples showed higher contents of nickel, cobalt, molybdenum, and lead compared to the reference trees. Also, the pine needles had elevated lead levels, although by a lower factor. Birch leaves revealed surprising patterns of elevated element bioaccumulation factors, with barium reaching up to eight, offering the possibility to limit analyses to plant material for risk assessments instead of soil analysis.

Keywords: bioaccumulation; Lidköping (Sweden); metal uptake; porcelain brownfield; Scots pine; silver birch

1. Introduction

The western harbor in Lidköping, the so-called Hamnstaden, is an industrial area nowadays also with a sewage treatment plant and a small marina. In the 20th century, its northern part was used as landfill for the waste from the surrounding industries, such as a porcelain factory [1]. Due to plans for reusing this brownfield for residential areas, risk assessments including soil and ground water analysis alongside bioavailability studies are needed and ongoing. The soil-polluting elements arsenic, cadmium, chromium, copper, lead, nickel, and zinc are of particular concern for residents via different uptake routes, such as soil inhalation, dermal contact, and direct/indirect ingestion [2].

The porcelain industry in Norra Hamnstaden started in 1892 with the painting of imported porcelain [3]. In 1912, the production of feldspar-based porcelain started [3] which is pre-fired at 900 °C, glazed, and finally fired at 1400 °C. The starting materials mainly consist of alumina and iron silicates, but also kaolinite, illite, and chlorite [4]. Furthermore, the clay composition contains various impurities, such as quartz, calcium and magnesium carbonates, alkali and alkaline earth metal oxides, and different iron compounds, which

https://doi.org/10.3390/molecules30102196

differ a lot even when taken from the same site [4]. Nevertheless, the ratio of certain trace elements can be used for locating production sites of archeological porcelain findings [5]. Another contribution to the elemental composition of such products is the glazing of the porcelain surface by applying a thin outer glass layer, its main constituent being silica (SiO_2). To lower the melting point and to improve the glaze's properties, different metal oxides are added, e.g., white lead ($2PbCO_3{*}Pb(OH)_2$) [4,6]. Various metal salts are added to the glaze for coloring, mainly containing the following eight metals, titanium, vanadium, chromium, manganese, iron, cobalt, nickel, and copper, but in rare cases also compounds of gold, uranium, cadmium, antimony, and selenium are used [4]. Leaching of coloring salts from properly prepared glazes is not of concern [7], but porcelain waste stored in the environment is supposed to undergo degradation processes which might lead to leaching of potentially toxic elements to the surroundings [8]. Furthermore, it has been found that aging of contaminated soil increases the bioavailability of certain elements, e.g., lead [9,10]. Consequently, the number of the potentially polluting elements as well as their contents are expected to vary widely within the deposit and need to be investigated as the basis for site-specific risk assessment. The present investigation focused on the bioavailability of selected elements via monitoring of tree leaves. Besides foliar uptake, metals, nonmetals, as well as metalloids enter plants via roots, thus the elemental pattern from the soil is reflected in these plant parts [11–13]. Whilst essential elements need to be taken up for plant physiology and to avoid deficiencies [14,15], this uptake might be interfered by non-essential, even potentially toxic elements using similar uptake and storage pathways. During the vegetation period, harmful elements are accumulating in leaves so that they can be removed in autumn [16]. Metal uptake from soil into plants depends on various factors, one being the metal load in the surroundings, i.e., the soil composition and the pollution of the respective site [17–20]. Furthermore, the characteristic climatic conditions of each site which even changes from year to year [16] determine the obtained elemental contents in a plant tissue. Apart from these influencing parameters, analyzing various plants grown in the same area has shown that the uptake and accumulation of elements is species-specific [21,22]. In addition to these influencing parameters, the plants' strategies to minimize either the uptake of potentially toxic elements or, later on, toxicity by these elements, are manifold [23–25], thus making the overall interpretation more complex.

The area of interest has already been studied, but these previous investigations carried out between 1999 and 2013 were mainly focused on total soil contents of certain contaminants [1], but for local risk assessment studies the actual bioavailability, e.g., via plant uptake, is needed. Thus, for the present investigation, two species were chosen, namely a deciduous tree birch (*Betula pendula*) as well as a conifer Scots pine (*Pinus sylvestris*) which both grow on the area of interest, i.e., the old porcelain deposit (see Figure 1). Birches are medium-sized deciduous trees found in Europe, parts of Asia, but also in America. As a pioneer species it grows after a forest fire or on bare land among the first tree species [26]. Scots pines are the most widely distributed pine tree species and used for monitoring environmental pollution. Like birches, Scots pines are pioneer species that can colonize nutrient-poor soils, e.g., contaminated sites [26,27]. Another advantage of this tree is that needles can reach a higher age than birch leaves, thus the accumulation can be studied over a longer period [22,28].

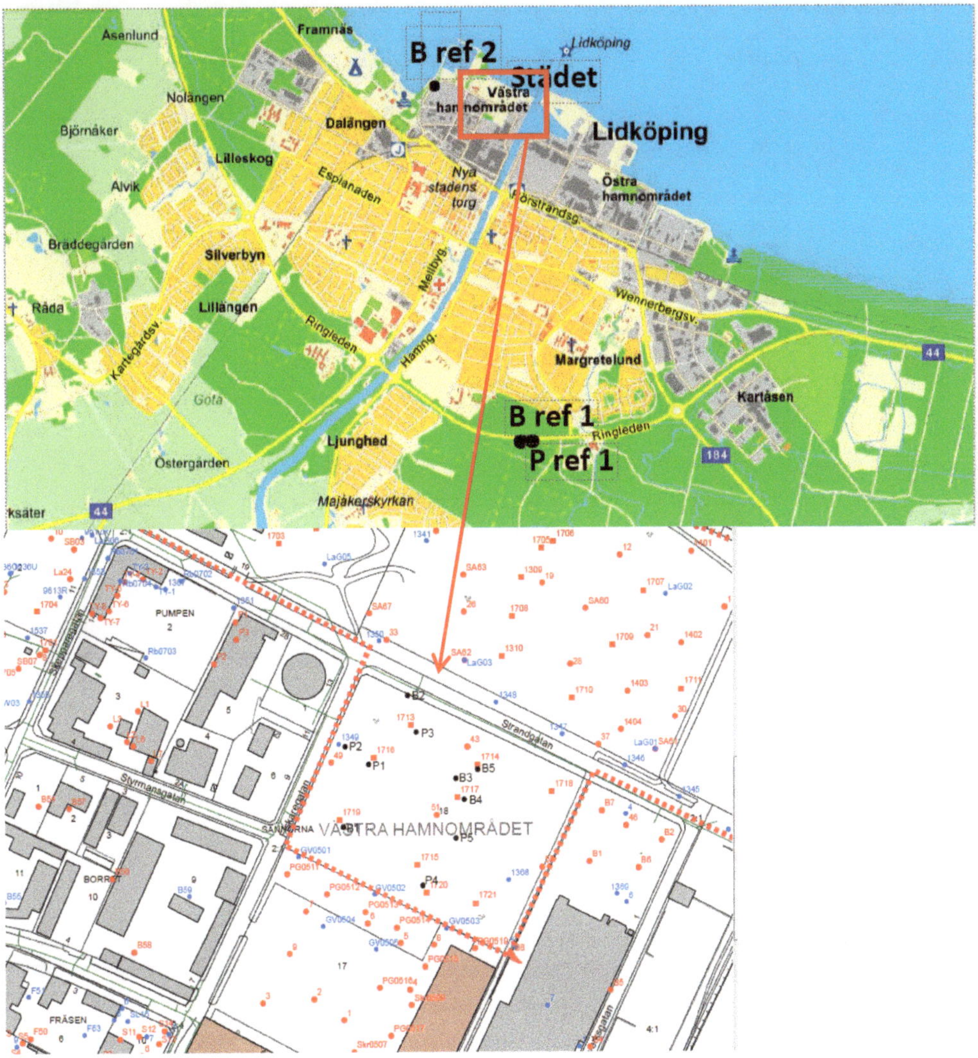

Figure 1. Study area and tree sampling plan for reference trees (BR1, BR2, PR1) and contaminated area (birches B1–B5, pines P1–P5) [graphics provided by Jordnära Miljökonsult].

2. Results and Discussion

2.1. Figures of Merit of the Analytical Procedure

The applicability of the chosen method for the given analysis was proven by the results obtained for the analyzed standard reference material. Accuracy expressed by trueness (as recovery in %) and precision (as RSD in %) for all analytes obtained by analyzing the pine needle standard reference material NIST 1575a is given in Table 1. Overall, the recoveries for the elements with certified as well as with reference mass fractions were in the range of 86% up to 105%, their relative standard deviations being all below 3%. All external calibration curves had R^2 values beyond 0.997.

Table 1. Recoveries and RSD for all analytes determined by SRM analysis.

Analyte	Recovery (%)	RSD (%)
As	101	2.9
Ba	86.2	2.3
Cd	98.9	1.4
Co	85.9	2.6
Cu	96.0	1.5
Cr	105	2.9
Mn	90.9	1.8
Mo	n/a *	n/a
Ni	96.1	2.5
Pb	91.2	2.8
V	n/a *	n/a
Zn	88.4	2.7

* not applicable, since not certified in the pine needle SRM.

2.2. Elemental Contents in Birch Leaves

The results obtained for the birch leaves from the contaminated site along with those for the two reference trees are shown in the following charts. For each tree, the determined elemental contents are given as mean values in mg/kg dried leaf. All mass fractions obtained for the trees grown in the contaminated area are compared with those for the reference trees using a *t*-test based on 95% probability level; all calculated *p*-values are shown in Table 2. Thus, all values <0.05 stand for statistically significant differences. Whilst Pb and Co differ for all trees from the brownfield, none up to three differ for the other elements.

Table 2. *p*-values for *t*-tests between each birch tree (B1–B5) and the reference trees (BR1 + BR2); statistically significant differences are marked bold ($p < 0.05$).

Analyte ↓/Tree →	B1	B2	B3	B4	B5
As	**0.030**	0.33	0.30	**0.0054**	0.38
Ba	**1.0×10^{-6}**	0.10	**4.6×10^{-7}**	0.097	**0.00011**
Cd	0.62	0.19	0.52	0.064	0.67
Co	**0.00025**	**7.9×10^{-7}**	**3.5×10^{-9}**	**2.0×10^{-5}**	**2.9×10^{-7}**
Cu	0.12	0.87	**0.0022**	**0.039**	0.85
Cr	0.29	0.61	0.23	0.61	0.32
Mo	0.34	**1.3×10^{-7}**	**0.0056**	**0.0036**	**0.018**
Ni	**0.00021**	0.70	0.76	0.25	**0.0025**
Pb	**2.7×10^{-5}**	**8.4×10^{-8}**	**1.5×10^{-6}**	**1.6×10^{-6}**	**5.1×10^{-6}**
V	**0.00058**	0.68	0.77	0.13	0.32
Zn	**8.7×10^{-5}**	0.41	**0.0095**	**0.0018**	**0.0029**

Lead was detected in soil samples from the area of interest in contents up to 52×10^3 mg/kg [29]. Birches are not considered as hyper-accumulators regarding Pb, i.e., accumulating more than 1×10^3 mg/kg of this element in the leaves [30]. Nevertheless, statistically significantly elevated mass fractions were found in all five trees from the contaminated site (see Figure 2). The Pb levels rise during the vegetation period since trees tend to get rid of toxic elements in fall by loss of the leaves. Even if all trees from the contaminated site accumulated more Pb than the reference trees, no direct correlation to the soil content of this metal or the age/stem diameter of the trees could be found. Birches B1 and B3 had similar stem diameters. Sample B1, however, collected close to the soil sample with the highest Pb content, did not exhibit the highest content in the leaves, even containing less than B3 located in a less contaminated part of the site (Pb up to 7×10^3 mg/kg). Furthermore, tree B4, located close to B3 and with greater stem diameter, took up less Pb. This site-specific uptake can be explained by the fact that the metal uptake is not limited to one soil sampling spot, but the roots of trees are widespread and grow more towards less contaminated areas to avoid high uptake of harmful elements [31].

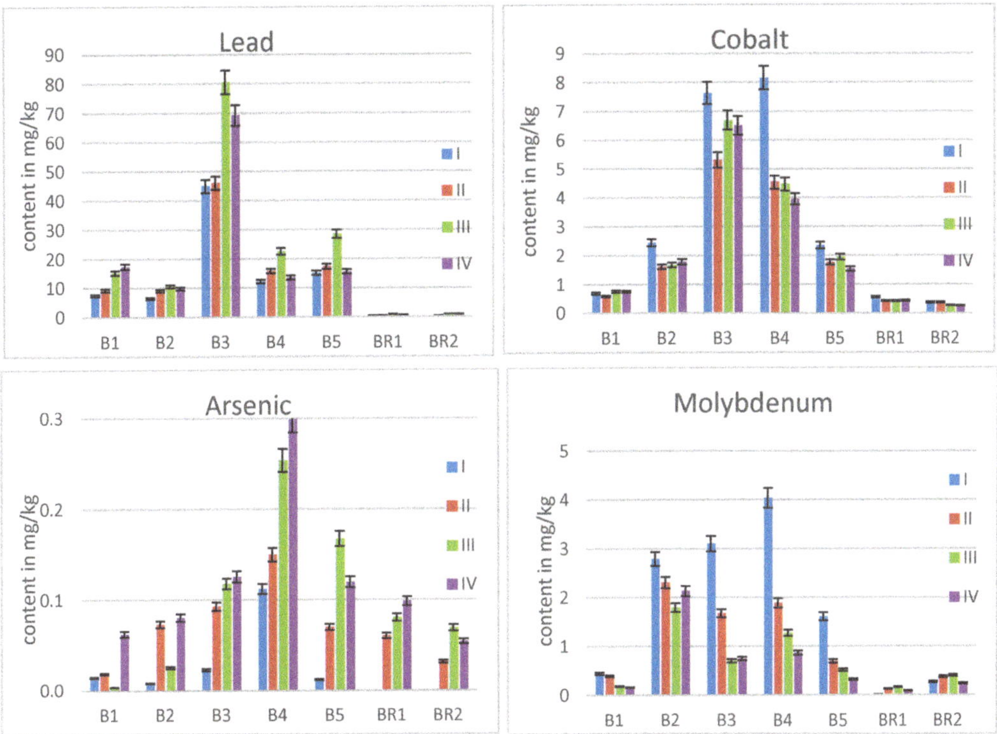

Figure 2. Mean contents of Pb, Co, As, and Mo in birch leaves for all sampling times.

Also, for cobalt, higher uptake can be stated for trees grown on the contaminated soil (see Figure 2). However, in contrast to Pb, Co is essential for plants, being required by various enzymes [32]. The soil Co contents of the sampling site were, in decreasing order, 380 mg/kg (for B2) > 89 mg/kg (for B5) > 60 mg/kg (for B1) > 43 mg/kg (for B3 and B4) [1,29]. Looking at the order of the Co contents in the respective leaves, this order is not reflected, which is determined to be B3, B4 > B2, B5 > B1.

For arsenic, the soil contents did not vary a lot, but all data from the respective sampling spots were in the range from 6 mg/kg to 10 mg/kg. Nevertheless, the contents in the birch leaves were not that evenly distributed; B4 in particular had accumulated more As than the other trees (see Figure 2). Leaching tests on porcelain waste showed that no harmful effects are expected to the surrounding soil, surface water, and groundwater environment [33]; however, elevated soil As contents are stated for the study area [29] and also reflected in the leaves.

The findings for molybdenum are also shown in Figure 2. Regarding this element, an accumulation can be stated for all trees grown in the contaminated area, except for B1, whose Mo level was in the range of that found for the reference trees. The trend of decreasing Mo content with time was registered for all trees B1 to B5 in contrast to no significant change during the vegetation period in the leaves of the reference trees. This accumulation behavior is typical for essential micronutrients, which Mo is [34], but particularly obvious when more of the respective essential element is present.

A diverse picture was obtained for nickel (Figure 3). Whereas the soil Ni contents were all found to be in a comparable range, namely from 19 mg/kg to 28 mg/kg [1,29], the contents in the leaves varied widely. B1 and B5 contained statistically significant more Ni than B2, B3, and B4 whose Ni levels were in the range of the reference trees.

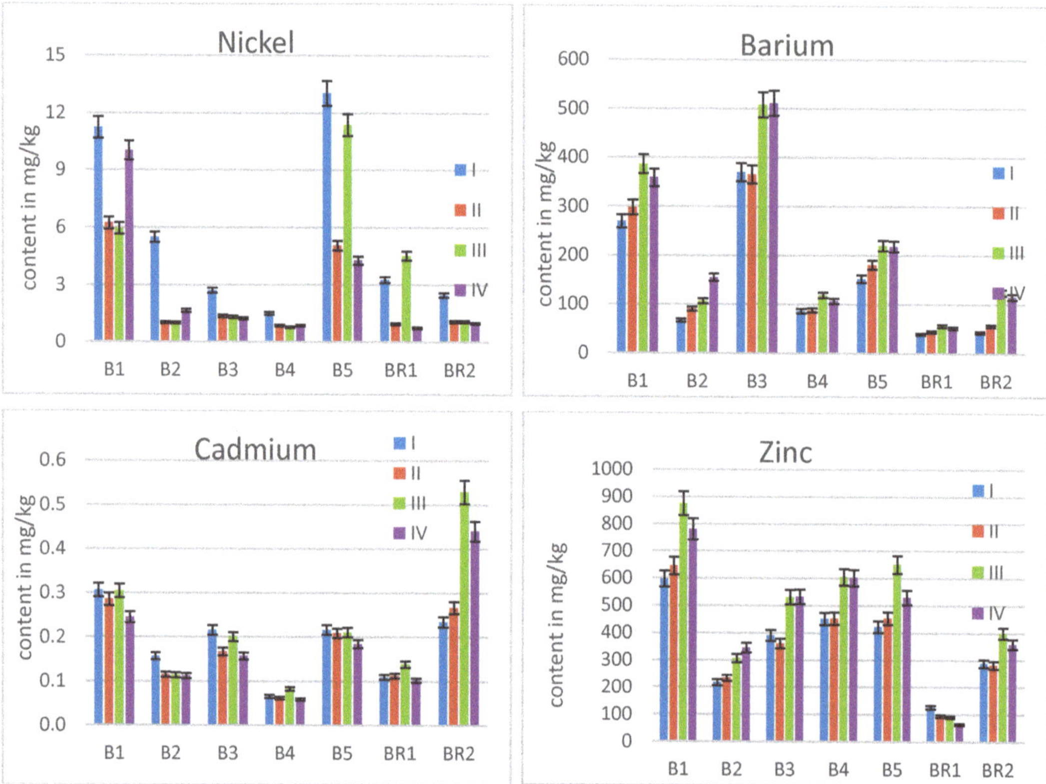

Figure 3. Mean contents of Ni, Ba, Cd, and Zn in birch leaves for all sampling times

Elevated contents for barium (Figure 3) were found in all sampled trees from the contaminated area, whereby the leaf contents for B2 and B4 were close to those of the reference trees. In all cases, an increase in time was found as expected for a non-essential element. Ba is the only element with an accumulation factor (content in leave/content in soil) beyond 1 for all trees B1 to B5 (Table 3). This ratio is sometimes also called soil-to-plant transfer factor [35].

Cadmium, a toxic metal to plants, was found in higher amounts in the leaves of the reference trees than in those from the contaminated site (compare Figure 3), the order of the contents being B4 < BR1, B2 < B3, B5 < B1 < BR2. Looking at the accumulation factors, values up to approx. 1 are determined for Cd (Table 3).

The results for Zn show that this metal was present in significantly lower amounts in the leaves of BR1 than in B1 to B5, but the values determined for BR2 were in the same order of magnitude as B2 to B5 (Figure 3). Only B1, showing the highest contents at all sampling times, differs from both reference trees significantly. Accumulation of elements strongly depends on the surroundings and characteristics of the site. Sitko and colleagues found site-specific elevations for Mn or Zn in birch leaves at different post-industrial heaps in Poland [36]. Dresler and colleagues investigated silver birch trees near Zn-Pb ore-processing sites, finding higher accumulation of trace metals in leaves, which further affects the accumulation of phenolic acids and flavonoids [37]. The dust in the smelter vicinity is rich in heavy metals, thus not only root uptake of the elements of interest is to be expected, as at the present study location of the porcelain brownfield.

Table 3. Accumulation factors for birch leaves (bold values are > 1) based on elemental contents in the leaves collected in July and one to two soil sampling sites close to the tree-growing area (soil sampling site 1719 for B1; 1713 for B2; 1717 for B3a and B4; 1714 for B3b and B5).

Analyte ↓/Tree→	B1	B2	B3a	B3b	B4	B5	Min	Max	Mean
As	0.0007	0.0027	0.0198	0.0206	0.0428	0.0293	0.0007	0.0428	0.0193
Ba	**2.9171**	**6.6390**	**7.8302**	**3.6299**	**1.8243**	**1.5727**	**1.5727**	**7.8302**	**4.0689**
Cd	0.6614	0.5701	0.2255	**1.0033**	0.0936	**1.0532**	0.0936	**1.0532**	0.6012
Co	0.0126	0.0044	0.1574	0.0752	0.1050	0.0219	0.0044	0.1574	0.0627
Cu	0.0325	0.1159	0.0419	0.0635	0.0492	0.1147	0.0325	0.1159	0.0696
Cr	0.0423	0.0453	0.0135	0.0223	0.0128	0.0094	0.0094	0.0453	0.0243
Mo	0.3924	n/a *	0.3763	n/a *	0.6851	n/a *	0.3763	0.6851	0.4846
Ni	0.2631	0.0527	0.0466	0.0684	0.0274	0.5987	0.0274	0.5987	0.1762
Pb	0.0008	0.0015	0.0160	0.0136	0.0045	0.0048	0.0008	0.0160	0.0069
V	0.0224	0.0222	0.0029	0.0076	0.0019	0.0134	0.0019	0.0224	0.0117
Zn	**1.2641**	0.7482	0.0772	0.6480	0.0879	0.7940	0.0772	**1.2641**	0.6032

n/a *: not applicable, since no soil Mo data available.

In contrast to the above discussed elements, no accumulation trends in the leaves from B1 to B5 in comparison to BR1 and BR2 were found for copper, vanadium, and chromium.

Based on the contents in the collected birch leaves, the respective accumulation factors were calculated for As, Ba, Co, Cu, Cr, Ni, Pb, V, and Zn for the samples collected in July which are presented in Table 3. The soil contents for the elements in a depth of 0.5 m to 1 m of the nearest soil sampling spot(s) to each sampled tree were used. These factors can only provide an estimation of the metal uptake since the roots of the trees are further distributed down than the depth of 1 m and not only close to the soil sampling site, which is even explicitly shown by B3 having a different soil sampling spot in the close vicinity. To avoid a high uptake of harmful elements, plants produce more and longer roots to more favorable regions [31] alongside other strategies to combat (heavy) metal toxicity, such as restricting metals to cell walls, vascular sequestration, and synthesizing biochemical compounds to minimize toxic effects [23]. This wide range of possibilities in handling toxic elements by plants explains the variations in the calculated accumulation factors and clearly show that the soil composition is only one parameter in the whole uptake and storage mechanism. Kozlov and coworkers found in their study on uptake roots of heavy metals by birch in an industrially polluted area that nickel, but not copper, effectively translocates from roots to shoots and leaves in mountain birch, with most contamination occurring due to dust particles on leaf surfaces [24]. This is reflected in the accumulation factors obtained in the present study for those elements listed in Table 3. The plant–soil relationship of various metals was investigated by Pająk and coworkers stating that Scots pine needles and silver birch and leaves show increased accumulation of Zn and Pb in response to Pb–Zn ore mining and processing plants, but not Cu, Cd, and Cr [26].

2.3. Elemental Contents in Scots Pine Needles

The pine needles were analyzed for the same elements as the birch leaves, and even when growing on the same contaminated area, the uptake and accumulation behavior is quite different, underlining the species specificity of these processes as already stated in literature [19]. The calculated p-values for each element determined in the fresh needles are shown in Table 4, to allow comparability with the birch leaves which only last one vegetation period. Lead, even if found in higher amounts than in the reference needles, does not differ significantly in any of the trees after one summer. Like for the birch leaves, Mo is accumulated to a significant extent in almost all sampled pine needles.

Table 4. *p*-values for *t*-tests between fresh needles of each pine tree (P1–P5) and the reference trees (PR1 + PR2); statistically significant differences are marked bold ($p < 0.05$).

Analyte ↓/Tree→	P1	P2	P3	P4	P5
As	0.12	0.99	0.35	0.83	0.50
Ba	0.87	0.52	0.56	$\mathbf{1.5 \times 10^{-5}}$	0.51
Cd	0.13	0.86	$\mathbf{2.4 \times 10^{-5}}$	**0.0022**	0.23
Co	**0.0012**	**0.026**	0.13	**0.0030**	0.13
Cu	0.96	0.27	0.44	0.66	0.30
Cr	0.094	0.95	0.76	0.67	0.14
Mo	**0.010**	**0.0016**	**0.00026**	**0.0030**	**0.0012**
Ni	0.60	0.38	0.37	**0.0024**	0.096
Pb	0.054	0.49	0.46	0.066	0.23
V	0.61	0.51	0.49	0.50	0.85
Zn	0.56	0.18	0.16	0.69	0.11

In contrast to deciduous trees, conifers offer the possibility to investigate long-term metal accumulation, since the needles of certain species can reach an age up to seven years. This fact was used in the present study by sampling and analyzing needle samples of three different ages. Even if no statistically significant differences in the mass fractions for Pb were found for the fresh roots, the 1-a and 2-a-old needles accumulated enough of this metal to differ significantly from the needles from the reference trees. The *p*-values for the *t*-tests are 0.0040, 0.19, 0.013, 5.0×10^{-7}, and 0.038 for the 1-a-old needles for P1, P2, P3, P4, and P5, respectively, and 0.00042, 0.099, 0.0042, 6.2×10^{-8}, and 0.0018 for the 2-a-old needles.

As already seen for the birch leaves, lead, as one of the main pollutants of interest, was found in the trees growing at the contaminated site in higher contents than in the reference tree(s), even if the difference is not as obvious as for the birch leaves. This harmful metal is accumulated in the needles over time, as can be seen from the values increasing with needle age (compare Figure 4). The determined Pb mass fractions for P4 are close to the data of P1 for the fresh shoots, but much higher for the one-year-old and even older needles. P4 grows close to the soil sampling spot with the lowest Pb soil levels (20 mg/kg to 280 mg/kg; [29]) compared to the other pine sampling places (Pb ranging from 1.3×10^3 mg/kg up to 52×10^3 mg/kg; [1,29]). This finding of the highest Pb levels in P4 seems to be of interest regarding metal uptake from soil. Even if the needles were washed prior to sample preparation to avoid metal contamination from precipitation, there is always a certain percentage of pollutant taken up via the stomata as well as via cuticular cracks [11]. A study of Scots pine needles in a mining area found a strong plant–soil relationship for Pb [26]. Çomaklı and colleague published that Scots pine needles are the plant tissue with the highest accumulation factors for many metals [38].

Not only for Pb, but also for other metals, namely Co, Mn, and Ba, P4 shows the highest contents, followed by P1. For illustration, the results for these three analytes are given in Figure 5 for the sampling time August (IV). In contrast, Cd was found in statistically higher contents only in P3 and P4. Regarding historical municipal solid waste landfill soils in Bratislava, Slovakia, which show high metal (loid) concentrations, Cd was determined as being the most bioaccessible [39].

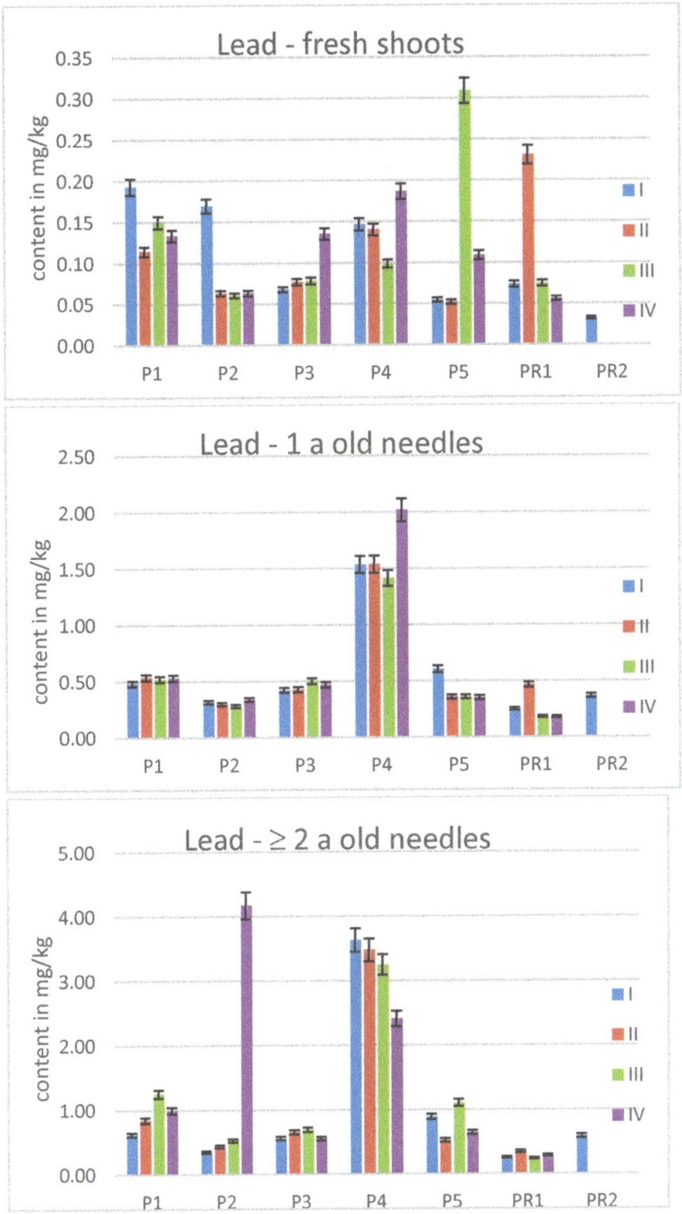

Figure 4. Lead content in pine needles of different ages for all sampling times.

Whereas the results for As and Zn differed in the birch leaves; no significant differences between pines from the contaminated and reference sites were found. In general, the obtained data for the pine needles are in the same order of magnitude than those for various pine species grown in the Lisičine Arboretum in Croatia [22].

Soil composition is an important parameter influencing the uptake and accumulation of (toxic) metals, but plants become tolerant through various strategies, like a mineral toxicity fence. It has been found that magnesium and iron levels in Scots pine needles play a crucial role in regulating heavy metal fluxes, which contribute to their survival and response to pollution effects [40].

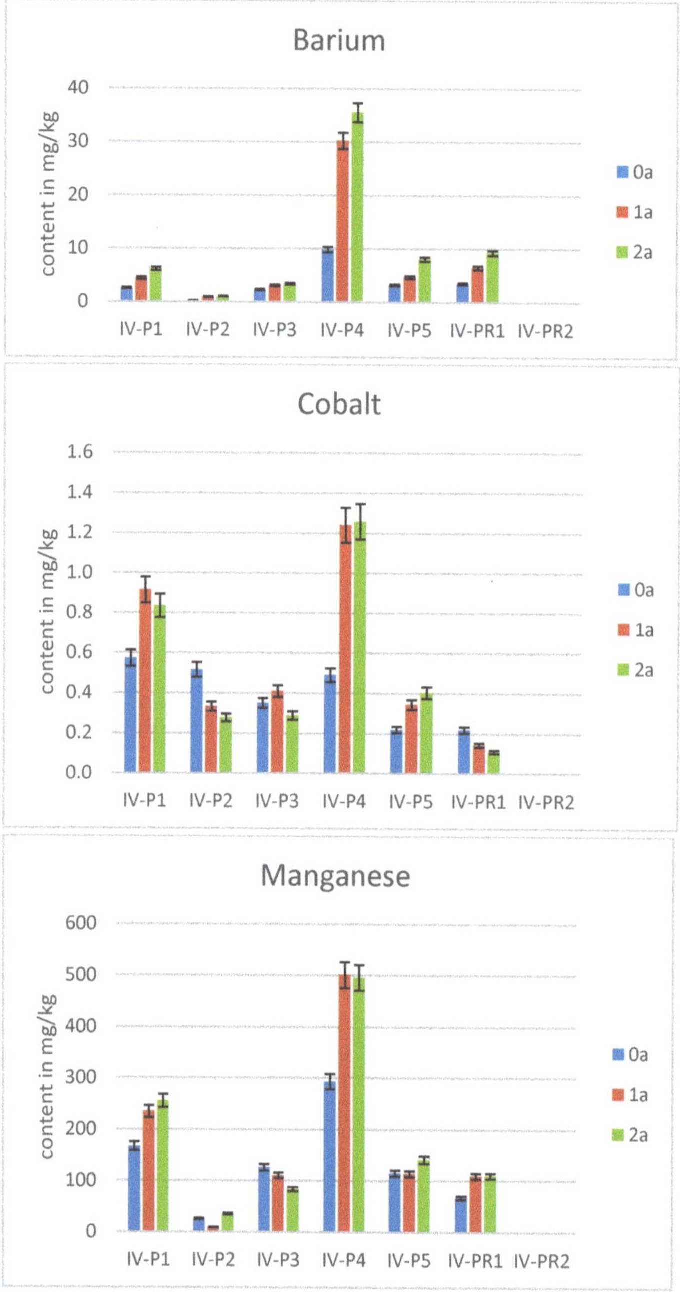

Figure 5. Barium, cobalt, and manganese in pine needles sampled in August.

2.4. Comparison of Birches and Scots Pines

As already pointed out in the respective subchapters on birch leaves and Scots pine needles, both show different uptake and accumulation behavior of the elements studied.

Not only the absolute mass fractions found, but also the comparison to the reference trees reflect this fact. Regarding Pb, up to 80 mg/kg was found for birches for B3, with the other trees showing leaf Pb mass fractions around 6 mg/kg to 30 mg/kg, whilst the needles contain much less after one, two, or even three and more vegetation periods, the maximum being around 4 mg/kg. Both birches and pines are used as bioindicators for metal contamination, but they differ in their mechanisms and efficiency of metal uptake and translocation. Birches tend to accumulate metals more from surface deposition and show distinct root-to-shoot translocation patterns, while pines primarily reflect soil metal concentrations and exhibit species-specific internal distribution. As shown by Kozlov et al., a significant portion (>80%) of nickel and copper in birch foliage comes from dust deposition on leaf surfaces, with only a minor part removed by washing, indicating strong canopy uptake [24]. For pine needles, surface deposition is less emphasized compared to birches [26]. In addition to the primary uptake root, internal translocation and distribution determines the mass fractions determined in the leaf tissues, the shoot-to-translocation being element-specific. Nickel is more effectively translocated from roots to shoots than copper, showing selective internal movement [24]. Internal transport of metals can be investigated using radionuclides like ^{90}Sr and ^{137}Cs; birches as well as pine show age-dependent internal distribution [41]. For detailed assessment of metal transfer and translocation in plants, so-called dynamic factors considering internal (physiological) as well as external (ecological) factors is recommended [42].

3. Materials and Methods

3.1. Sampling

Sampling was performed by a representative of Jordnära Miljökonsult once per month in the vegetation period from May to August 2018. In the following, these sampling times are indicated as I—May; II—June; III—July, and IV—August. As recommended by the ICP Forest Manual on sampling leaves and needles [43], five trees per species growing in the contaminated area were sampled (B1–B5, P1–P5) along with two reference trees (BR1, BR2, PR1, PR2). For *Pinus sylvestris*, the second reference tree was cut after the first sampling, thus later needles were collected only from one reference tree. The exact sampling locations are shown in Figure 1. From each birch tree, 15–20 leaves were collected each month. Each pine tree was sampled for fresh needles, one-year-old needles, and ≥two-year-old needles. Leaves and needles were pooled by sampling occasion, tree, and age prior to further treatment.

3.2. Sample Preparation and Elemental Analysis

To avoid falsification of the data by metals attached to the leaf or needle surface, all samples were rinsed three times with ultrapure water (18.2 MΩ) prior to the drying step at 105 °C. The dried plant material was then ground and homogenized. Three portions of each sample (approx. 100 mg, weighed to the nearest 0.1 mg) underwent acidic microwave-assisted digestion using nitric acid (65% *w*/*w* originating from Merck, Darmstadt, Germany) and hydrogen peroxide (30% *w*/*w* originating from Supelco, Darmstadt, Germany), with the following volumes used: for birch leaves 6 mL concentrated nitric acid, 3 mL deionized ultrapure water, and 1 mL hydrogen peroxide solution; for pine needles 6 mL concentrated nitric acid, 3 mL deionized ultrapure water, and 2 mL hydrogen peroxide solution. The instrument used was a CEM MarsV microwave oven. The temperature program included two steps with 20 min each (heating to 150 °C and 180 °C, respectively). The obtained clear solutions were filled up to 50 mL with ultrapure water (18.2 MΩ) from the own in-house water-purification facility. Prior to the measurement, each solution was filtered through a

0.2 μm polypropylene filter and the internal standard added (Rh with a final concentration of 10 μg/L). Blank solutions were prepared analogously.

For quality assurance (repeatability, trueness), six aliquots of the standard reference material SRM 1575a—Trace Elements in Pine Needles (Gaithersburg, MD, USA) were treated in the same way as the samples.

All obtained digest solutions including blanks were analyzed using inductively coupled plasma mass spectrometry (ICP-MS; Agilent 7500cx) after further dilution of 1:10 using 1% *w/w* nitric acid. The instrumental conditions applied are listed in Table 5. External calibration based on five multi-elemental standard solutions prepared from ICP multi-element standard solution VI (Merck, Darmstadt, Germany) was used for quantification.

Table 5. Instrumental conditions for analytical method used.

Parameter	ICP-MS
Instrument	Agilent 7500cx ICP-MS (Agilent, Tokyo, Japan)
Output power	1500 W
Argon flows	Plasma:15 L min^{-1} Auxiliary: 0.9 L min^{-1} Nebulizer: 0.2 L min^{-1}
Sample flow	0.3 mL min^{-1}
Nebulizer	MicroMist
Spray chamber	Scott double pass
Isotopes	^{51}V, ^{53}Cr, ^{55}Mn, ^{59}Co, ^{60}Ni, ^{63}Cu, ^{66}Zn, ^{75}As, ^{95}Mo,^{111}Cd, ^{137}Ba, $^{204+206+207+208}Pb$
Collison cell	Off: Ba, Pb, Cd, Co, Cu, Mn, Mo, Ni, Zn On (He 5 mL/min): V, Cr, As

3.3. *Data Evaluation and Statistical Tests*

All raw data from the instrument (mass concentration of the analytes) were converted into contents in mg/kg considering digestion blank, final volume, dilution step, and mass of sample digested. The mean and standard deviation was calculated for each sample based on the triplicate digestion. Microsoft Office Excel, version 2016 was used to perform the calculations as well as statistical tests. *t*-tests were applied to check for statistical significance of differences between elemental content determined in leaves from contaminated site vs. those from reference trees (based on $p = 0.05$ as limit). Accumulation factor calculations are based on the soil sampling site closest to the respective tree; in the case of B3, two spots were similar in distance, thus both were considered. The formula used was content of element in plant material divided by the total content of the element in the respective soil sample, whereby the plant content is always calculated on a dry matter basis.

4. Conclusions

Based on the results obtained for birch leaves and pine needles sampled in the contaminated site at the porcelain deposit in Lidköping (Sweden), elevated bioavailability compared to reference sites can be stated. Differences in the accumulation tendencies in leaves were found for As, Ba, Pb, Co, Cu, Cr, Ni, V, and Zn. Furthermore, different uptake patterns were observed for birch leaves and pine needles, whereby the latter also show a variation with increasing needle age. These differences highlight the importance of species selection and understanding uptake pathways when using trees as bioindicators of metal pollution. Comparison of the elemental contents in the plant tissues and the respective soil samples showed the limited significance of a direct correlation to soil data due to wide distribution of roots in the soil along with strategies of the plants to avoid uptake of harmful elements. The accumulation factors are mainly below 1; the highest values (up to 8) were obtained for Ba.

Author Contributions: Conceptualization, H.O., V.S. and M.Z.; methodology, H.O., V.S. and M.Z.; validation, H.O., V.S. and M.Z.; investigation, H.O., V.S. and M.Z.; resources, H.O.; data curation, H.O., V.S. and M.Z.; writing—original draft preparation, H.O., V.S. and M.Z.; writing—review and editing, M.Z.; visualization, M.Z. All authors have read and agreed to the published version of the manuscript.

Funding: This research received no external funding.

Institutional Review Board Statement: Not applicable.

Informed Consent Statement: Not applicable.

Data Availability Statement: All data will be available on reasonable request.

Acknowledgments: The authors thank all assisting persons in experimental work and the university for the possibility to use lab-facilities and instruments, and Jordnära Miljökonsult for providing the samples.

Conflicts of Interest: The authors declare no conflicts of interest.

References

1. WSP. *PM Historik Och Tidigare Verksamheter—Norra Delen av Hamnstaden Lidköping*; WSP: Gothenburg, Sweden, 2014.
2. Jiang, Y.; Sihong, C.; Jianwei, L.; Yue, Y.; Yanjiao, C.; Aichen, Z.; Hongbin, C. Source apportionment and health risk assessment of heavy metals in soil for a township in Jiangsu Province, China. *Chemosphere* **2017**, *168*, 1658–1668. [CrossRef] [PubMed]
3. Rörstrand Museum. n.d. ALP Lidköping. Available online: https://rorstrand-museum.se// (accessed on 1 August 2022).
4. Denio, A. Chemistry for potters. *J. Chem. Educ.* **1980**, *57*, 272–275. [CrossRef]
5. Tabata, M.; Yagi, N.; Nishimoto, J.; Ghaffar, A. Estimation of places of production of porcelains of unknown origins excavated at the Mietsu Naval Facility site based on differences in the solubility of trace metals during the elutriation process. *J. Archaeol. Sci. Rep.* **2021**, *36*, 102823. [CrossRef]
6. Tunstall, S.; Amarasiriwardena, D. Characterization of lead and lead leaching properties of lead glazed ceramics from the Solis Valley, Mexico, using inductively coupled plasma-mass spectrometry (ICP-MS) and diffuse reflectance infrared Fourier transform spectroscopy (DRIFT). *Microchem. J.* **2002**, *73*, 335–347. [CrossRef]
7. Mohamed, N.; Chin, Y.; Pok, F. Leaching of lead from local ceramic tableware. *Food Chem.* **1995**, *54*, 245–249. [CrossRef]
8. Frankel, G.; Vienna, J.; Lian, J.; Scully, J.; Gin, S.; Ryan, J.; Wang, J.; Kim, S.; Windl, W.; Du, J. A comparative review of the aqueous corrosion of glasses, crystalline ceramics, and metals. *NPJ Mater. Degrad.* **2018**, *2*, 15. [CrossRef]
9. Schreck, E.; Foucault, Y.; Geret, F.; Pradère, P.; Dumat, C. Influence of soil ageing on bioavailability and ecotoxicity of lead carried by process waste metallic ultrafine particles. *Chemosphere* **2011**, *85*, 1555–1562. [CrossRef] [PubMed]
10. Wijayawardena, M.A.A.; Naidu, R.; Megharaj, M.; Lamb, D.; Thavamani, P.; Kuchel, T. Influence of ageing on lead bioavailability in soils: A swine study. *Environ. Sci. Pollut. Res. Int.* **2015**, *22*, 8979–8988. [CrossRef] [PubMed]
11. Shahid, M.; Dumat, C.; Khalid, S.; Schreck, E.; Xiong, T.; Niazi, N.K. Foliar heavy metal uptake, toxicity and detoxification in plants: A comparison of foliar and root metal uptake. *J. Hazard. Mater.* **2017**, *325*, 36–58. [CrossRef]
12. Maurice, C.; Lagerkvist, A. Using *Betula pendula* and *Telephora caryophyllea* for Soil Pollution Assessment. *J. Soil. Contam.* **2000**, *9*, 31–50. [CrossRef]
13. Zeiner, M.; Juranović Cindrić, I. Accumulation of major, minor and trace elements in pine needles (*Pinus nigra*) in Vienna (Austria). *Molecules* **2021**, *26*, 3318. [CrossRef] [PubMed]
14. Soetan, K.O.; Olaiya, C.O.; Oyewole, O.E. The importance of mineral elements for humans, domestic animals and plants: A review. *Afr. J. Food Sci.* **2010**, *4*, 200–222. [CrossRef]
15. Steinnes, E. Soil and human health. In *Sustaining Soil Productivity in Response to Global Climate Change: Science, Policy, and Ethics*, 1st ed.; Sauer, T.J., Norman, J.M., Sivakumar, M.V.K., Eds.; Wiley-Blackwell: Oxford, UK, 2011.
16. Zeiner, M.; Juranović Cindrić, I.; Mihajlov Konanov, D.; Stingeder, G. Determination of selected toxic elements in leaves of White Hawthorn grown in a remote area. *E3S Web Conf.* **2013**, *1*, 34003. [CrossRef]

17. Serbula, S.M.; Kalinovic, T.S.; Ilic, A.A.; Kalinovic, J.V.; Steharnik, M.M. Assessment of Airborne Heavy Metal Pollution Using *Pinus* spp. and *Tilia* spp. *Aerosol Air Qual. Res.* **2013**, *13*, 563–573. [CrossRef]
18. Zeiner, M.; Juranović Cindrić, I.; Ivanović, M.; Medunic, G. Availability of Selected (Pollutant) Elements and their Influence on Soil Composition in Urban Area. *Croat. Chem. Acta* **2015**, *88*, 23–33. [CrossRef]
19. Zeiner, M.; Juranović Cindrić, I.; Pozgaj, M.; Pirkl, R.; Šilić, T.; Stingeder, G. Influence of soil composition on the major, minor and trace metal content of Velebit biomedical plants. *J. Pharm. Biomed. Anal.* **2015**, *106*, 153. [CrossRef]
20. Jurić, D.; Puntarić, D.; Gvozdić, V.; Vidosavljević, D.; Lonèarić, Z.; Puntarić, A.; Puntarić, E.; Puntarić, I.; Vidosavljević, M.; Begović, L.; et al. Cabbage (*Brassica oleracea* var. *capitata*) as possible indicator of wartime metal and metalloid contamination in eastern Croatia (ICP-MS method). *Acta Agric. Scand. B Soil. Plant Sci.* **2017**, *67*, 270–277. [CrossRef]
21. Juranović Cindrić, I.; Zeiner, M.; Starčević, A.; Liber, Z.; Rusak, G.; Idžojtić, M.; Stingeder, G. Influence of F1 hybridization on the metal uptake behaviour of pine trees (*Pinus nigra* x *Pinus thunbergiana; Pinus thunbergiana* x *Pinus nigra*). *J. Trace Elem. Med. Biol.* **2018**, *48*, 190–195. [CrossRef]
22. Juranović Cindrić, I.; Zeiner, M.; Starčević, A.; Stingeder, G. Metals in pine needles: Characterisation of bio-indicators depending on species. *Int. J. Env. Sci. Technol.* **2019**, *16*, 4339–4346. [CrossRef]
23. Ejaz, U.; Khan, S.; Khalid, N.; Ahmad, Z.; Jehangir, S.; Rizvi, Z.; Lho, L.; Han, H.; Raposo, A. Detoxifying the heavy metals: A multipronged study of tolerance strategies against heavy metals toxicity in plants. *Front. Plant Sci.* **2003**, *14*, 1154571. [CrossRef]
24. Kozlov, M.; Haukioja, E.; Bakhtiarov, A.; Stroganov, D.; Zimina, S. Root versus canopy uptake of heavy metals by birch in an industrially polluted area: Contrasting behaviour of nickel and copper. *Environ. Pollut.* **2000**, *107*, 413–420. [CrossRef] [PubMed]
25. Gatasheh, M.; Abbas, T.; Shaffique, S.; Kang, S.; Lee, I.; Shah, A. Comparative analysis of biodiversity, physiology, and anatomical adaptations in riparian flora exposed to industrial pollution stress. *Sci. Rep.* **2025**, *15*, 3006. [CrossRef] [PubMed]
26. Pająk, M.; Halecki, W.; Gąsiorek, M. Accumulative response of Scots pine (*Pinus sylvestris* L.) and silver birch (*Betula pendula* Roth) to heavy metals enhanced by Pb-Zn ore mining and processing plants: Explicitly spatial considerations of ordinary kriging based on a GIS approach. *Chemosphere* **2017**, *168*, 851–859. [CrossRef] [PubMed]
27. Vacek, Z.; Linda, R.; Cukor, J.; Vacek, S.; Šimůnek, V.; Gallo, J.; Vančura, K. Scots pine (*Pinus sylvestris* L.), the suitable pioneer species for afforestation of reclamation sites? *For. Ecol. Manag.* **2021**, *485*, 118951. [CrossRef]
28. Parzych, A.; Mochnacký, S.; Sobisz, Z.; Kurhaluk, N.; Polláková, N. Accumulation of heavy metals in needles and bark of Pinus species. *Folia Pol. Ser. A* **2017**, *59*, 34–44. [CrossRef]
29. Hederfeld, G. Risk Estimation of Multi-Polluted Soils in Contact with Lacustrine Systems. Master's Thesis, Örebro University, Örebro, Sweden, 2018. Available online: https://www.diva-portal.org/smash/record.jsf?pid=diva2:1266488&dswid=9903 (accessed on 20 March 2025).
30. Baker, A.J.M.; Brooks, R.R. Terrestrial Higher Plants which Hyperaccumulate Metallic Elements—A Review of their Distribution. *Ecol. Phytochem. Biorecovery* **1989**, *1*, 81–126. [CrossRef]
31. Khare, D.; Mitsuda, N.; Lee, S.; Song, W.; Hwang, D.; Ohme-Takagi, M.; Martinoia, E.; Lee, Y.; Hwang, J. Root avoidance of toxic metals requires the GeBP-LIKE 4 transcription factor in Arabidopsis thaliana. *New Phytol.* **2017**, *213*, 1257–1273. [CrossRef]
32. Hu, X.; Wei, X.; Ling, J.; Chen, J. Cobalt: An Essential Micronutrient for Plant Growth? *Front. Plant Sci.* **2021**, *12*, 768523. [CrossRef]
33. Chen, Y.; Zhou, Y. The contents and release behavior of heavy metals in construction and demolition waste used in freeway construction. *Environ. Sci. Pollut. Res.* **2020**, *27*, 1078–1086. [CrossRef]
34. Kaiser, B.N.; Gridley, K.L.; Ngaire Brady, J.; Phillips, T.; Tyerman, S.D. The role of molybdenum in agricultural plant production. *Ann. Bot.* **2005**, *96*, 745–754. [CrossRef]
35. Intawongse, M.; Dean, J.R. Uptake of heavy metals by vegetable plants grown on contaminated soil and their bioavailability in the human gastrointestinal tract. *Food Addit. Contam.* **2006**, *23*, 36–48. [CrossRef] [PubMed]
36. Sitko, K.; Opała-Owczarek, M.; Jemioła, G.; Gieroń, Ż.; Szopiński, M.; Owczarek, P.; Rudnicka, M.; Małkowski, E. Effect of Drought and Heavy Metal Contamination on Growth and Photosynthesis of Silver Birch Trees Growing on Post-Industrial Heaps. *Cells* **2022**, *11*, 53. [CrossRef] [PubMed]
37. Dresler, S.; Wójciak, M.; Sowa, I.; Sawicki, J.; Strzemski, M.; Hawrylak-Nowak, B.; Hanaka, A. Accumulation ability of trace metals by silver birch leaves in areas contaminated by Zn-Pb ore processing: Effects of excessive trace metal accumulation on specialized metabolism. *Chemosphere* **2024**, *362*, 142719. [CrossRef] [PubMed]
38. Çomaklı, E.; Bingöl, M. Heavy metal accumulation of urban Scots pine (*Pinus sylvestris* L.) plantation. *Environ. Monit. Assess.* **2021**, *193*, 192. [CrossRef] [PubMed]
39. Hiller, E.; Faragó, T.; Kolesár, M.; Filová, L.; Mihaljevič, M.; Jurkovič, Ľ.; Demko, R.; Machlica, A.; Štefánek, J.; Vítková, M. Metal(loid)s in urban soil from historical municipal solid waste landfill: Geochemistry, source apportionment, bioaccessibility testing and human health risks. *Chemosphere* **2024**, *362*, 142677. [CrossRef]
40. Tatuśko-Krygier, N.; Diatta, J.; Chudzińska, E.; Waraczewska, Z.; Gawroński, D.; Youssef, N. Bioactive levels of Zn, Pb, Cu, Cd and Mg, Fe in pollution sensitive and tolerant Scots pines needles—Is survival mineral-dependent? *Ecol. Indic.* **2023**, *146*, 109751. [CrossRef]

41. Holiaka, D.; Yoschenko, V.; Cherniaiev, O.; Moskaliuk, A.; Lesnik, O.; Levchuk, S.; Holiaka, M.; Gumenuk, V.; Kovbasa, Y.; Borsuk, O.; et al. Variability of activity concentrations and radial distributions of 137Cs and 90Sr in trunk wood of Scots pine and Silver birch. *J. Environ. Radioact.* **2023**, *263*, 107186. [CrossRef]
42. Baltrėnaitė, E.; Lietuvninkas, A.; Baltrėnas, P. Use of Dynamic Factors to Assess Metal Uptake and Transfer in Plants—Example of Trees. *Water Air Soil. Pollut.* **2012**, *223*, 4297–4306. [CrossRef]
43. Rautio, P.; Fürst, A.; Stefan, K.; Raitio, H.; Bartels, U. Part XII: Sampling and Analysis of Needles and Leaves. In *Manual on Methods and Criteria for Harmonized Sampling, Assessment, Monitoring and Analysis of the Effects of Air Pollution on Forests*; UNECE ICP Forests Programme Co-ordinating Centre, Ed.; Thünen Institute of Forest Ecosystems: Eberswalde, Germany, 2016; p. 19. Available online: http://icp-forests.net/page/icp-forests-manual (accessed on 1 August 2022).

molecules

MDPI

Article

Optimization and Validation of a QuEChERS-Based Method Combined with Gas Chromatography–Tandem Mass Spectrometry for Analyzing Pesticide Residues in Edible Insect Samples

Phannika Tongchai [1,2], Nootchakarn Sawarng [3], Anurak Wongta [1,2], Udomsap Jaitham [1,2], Kunrunya Sutan [2], Saweang Kawichai [2], Chuleui Jung [4], Bajaree Chuttong [5,*] and Surat Hongsibsong [1,2,*]

1 School of Health Sciences Research, Research Institute for Health Sciences, Chiang Mai University, Chiang Mai 50200, Thailand; phannika_tongchai@cmu.ac.th (P.T.); anurak.wongta@cmu.ac.th (A.W.); udomsap_j@cmu.ac.th (U.J.)

2 Environmental, Occupational Health Sciences and NCD Center of Excellence, Research Institute for Health Sciences, Chiang Mai University, Chiang Mai 50200, Thailand; kunrunya.s@cmu.ac.th (K.S.); sawaeng.k@gmail.com (S.K.)

3 Faculty of Public Health, Chiang Rai Rajabhat University, Chiang Rai 57100, Thailand; nootchakarn.saw@crru.ac.th

4 Department of Plant Medicals, Gyeongkuk National University, Andong 36729, Republic of Korea; cjung@gknu.ac.kr

5 Meliponini and Apini Research Laboratory, Department of Entomology and Plant Pathology, Faculty of Agriculture, Chiang Mai University, Chiang Mai 50200, Thailand

* Correspondence: bajaree.c@cmu.ac.th (B.C.); surat.hongsibsong@cmu.ac.th (S.H.)

Abstract

The increasing popularity of edible insects as a sustainable food source necessitates stringent safety measures to monitor pesticide contamination. This study aimed to assess and enhance a QuEChERS-based extraction method coupled with gas chromatography–tandem mass spectrometry (GC-MS/MS) for the quantification of pesticide residues in edible insects (bamboo caterpillars, house crickets, silkworm pupae, giant water bugs, and grasshoppers) by combining multiple individual insect specimens into a single, homogenized sample—five replicates were tested. The method was optimized by evaluating various extraction parameters and showed strong linearity for all 47 target pesticides, with correlation coefficients (R^2) ranging from 0.9940 to 0.9999. The limits of detection (LODs) varied between 1 and 10 μg/kg, while the limits of quantification (LOQs) ranged from 10 to 15 μg/kg. Recovery studies conducted at three fortification levels (10, 100, and 500 μg/kg) revealed recoveries ranging from 64.54% to 122.12%, that over 97.87% of the pesticides exhibited satisfactory recoveries within the range of 70–120%, and relative standard deviations (RSDs) below 20%, between 1.86% and 6.02%. Matrix effects (%MEs) range from −33.01% to 24.04%, and to those that experienced no effect. More than 94% of the analytes showed minimal ion suppression or enhancement. These results conform to the SANTE guidelines for monitoring pesticide residues in edible insects, enhancing food safety standards and safeguarding consumer protection.

Keywords: pesticide residues; edible insects; QuEChERS; GC-MS/MS; food safety; method validation; matrix effects; extraction optimization

1. Introduction

The global shift towards sustainability and eco-friendly food options has generated increasing interest in edible insects as a viable alternative protein source [1]. These insects provide a rich source of high-quality protein and essential fatty acids, vitamins, and minerals. Their production is notably more resource-efficient than that of conventional livestock, requiring significantly less land, water, and feed, which contributes to a more sustainable food system [2]. In numerous regions, especially across Asia, Africa, and Latin America, consuming insects has a long-standing history. In Thailand, the consumption of insects is deeply ingrained in cultural and culinary practices. Commonly consumed species include house crickets (*Acheta domesticus*), grasshoppers (*Melanoplus foedus*), silkworm (*Bombyx mori*) pupae, bamboo caterpillars (*Omphisa fuscidentalis*), and diving beetle (*Cybister limbatus*) [3].

While edible insects offer potential benefits for food security and nutrition, they also pose food safety concerns due to the risk of pesticide contamination. Insects are exposed to pesticides both directly from treated areas and indirectly through their contaminated diet [4]. Furthermore, specimens gathered from their natural environments might be influenced by agricultural runoffs or the use of pesticides within their ecosystems [5]. Several studies have reported the presence of pesticide residues in edible insect samples, specifically in food products made from *A. domesticus* collected in Vietnam, with some residues belonging to pesticides that are either banned or restricted in the country [6]. Previous studies identified nine agrochemicals in six species of edible insects, including insecticides, herbicides, and fungicides, with some species exceeding maximum residue limits for certain chemicals. Notably, the Nigerian cricket (*Brachytrupes membranaceus*) contained detectable residues of several pesticides currently in use, although the levels of most residues remained below the established maximum residue limit of 0.01 mg/kg, indicating compliance with safety standards [7]. Moreover, a study analyzing various species of edible insects, including mealworms (*Tenebrio molitor*), crickets, silkworm pupae, and grasshoppers, found that some samples contained high microbial counts, particularly those that were raw or inadequately processed. Chemical analysis also revealed the presence of heavy metals, including lead, cadmium, and arsenic, although the levels remained within Canadian safety limits. Mycotoxins and pesticide residues were either undetectable or present at very low concentrations [8]. The study of pesticides in edible insects is complicated by the elevated levels of fat and protein in the matrix [9]. Chromatographic techniques for quantifying pesticide residue concentrations often necessitate comprehensive lipid removal before sample introduction into the system [10]. Consequently, the cleanup protocol must be refined to eliminate sample-derived matrices efficiently. Various extraction methods have been developed for this purpose, including solid-phase extraction (SPE), liquid–liquid extraction (LLE), solid-phase microextraction (SPME), and the QuEChERS method, which stands for quick, easy, cheap, effective, rugged, and safe [11].

The QuEChERS method is widely recognized for its versatility and effectiveness in extracting pesticide residues from various food matrices. Its advantages include high recovery rates, reduced solvent consumption, and suitability for high-throughput workflows. This method typically involves extraction with acetonitrile and a salt mixture to induce phase separation, followed by cleanup using dispersive solid-phase extraction (dSPE) with sorbents such as primary secondary amine (PSA) and anhydrous magnesium sulfate ($MgSO_4$). Additional sorbents, such as C18 or graphitized carbon black (GCB), are sometimes employed to enhance cleanup, particularly for complex matrices [12–15]. Given the variability in fat content and matrix complexity among edible insect species, adapting and optimizing the QuEChERS protocol is essential to ensure accurate and reliable pesticide residue analysis.

Pesticide residue extraction methods are often analyzed using chromatographic techniques, such as gas chromatography–mass spectrometry (GC-MS/MS) or liquid chromatography–mass spectrometry (LC-MS/MS). Recently, liquid chromatography paired with quadrupole time-of-flight mass spectrometry (LC-QTOF/MS) has emerged as a powerful approach for both non-targeted and targeted pesticide screening in complex food matrices, owing to its enhanced resolution, accuracy, and sensitivity [16]. Studies have demonstrated the successful application of QuEChERS, combined with LC-MS/MS or GC-MS/MS, to detect pesticides in high-fat or complex matrices, such as rice, vegetable oils, and mealworms [17,18]. Previous studies using liquid chromatography–tandem mass spectrometry (LC-MS/MS) were developed to analyze 353 pesticides in *Tenebrio molitor* larvae. The method achieved limits of quantitation ≤10 μg/kg, with over 90% recovery and negligible matrix effects for most pesticide analyses [19].

The purpose of this study is to develop and validate an optimized QuEChERS-based extraction method combined with GC-MS/MS for multiresidue pesticide analysis in edible insects. The method was applied to real insect samples to assess potential pesticide contamination, supporting food safety monitoring efforts for insects, which are increasingly recognized as sustainable future food sources.

2. Results and Discussion

2.1. Optimization of Sample Extraction Procedure

The extraction of pesticide residues from edible insect matrices presents a significant analytical challenge due to their high protein and lipid content. In particular, lipids are highly soluble in organic solvents, which facilitates their co-extraction. To ensure accurate analysis, it is essential to purify the sample using appropriate pretreatment procedures prior to gas chromatography (GC) analysis [17,19]. The QuEChERS method was modified to use the optimal solvent/sample ratio. A standard mixed solution of 47 selected pesticides (100 μg/mL) was spiked into edible insect samples (2.5, 5.0, and 10.0 g). Each sample was transferred to a 50 mL centrifuge tube, to which varying amounts of acetonitrile (ACN) (5.0, 10.0, and 15.0 mL) were added, followed by 5 mL of water. The mixtures were then agitated for 5 min. Subsequently, a QuEChERS extraction package comprising 6 g of magnesium sulfate ($MgSO_4$) and 1.5 g of sodium citrate ($Na_3C_6H_5O_7$) was incorporated to enhance phase separation and optimize cleanup efficacy.

The number of pesticides extracted increased significantly with the addition of higher volumes of ACN across all sample sizes, as illustrated in Figure 1. In the 2.5 g sample, the number of detectable pesticides increased markedly from 21 (extracted with 5 mL of ACN) to 45 (with 15 mL of ACN), indicating that extraction efficiency improves with increased solvent volume. The 5.0 g and 10.0 g samples exhibited a similar trend, though their overall recovery rates were lower. Using a larger volume of solvent relative to the sample size is especially critical for smaller samples, as it facilitates the separation of lipophilic pesticide residues into the extraction solvent [20–22]. This study demonstrated that increasing the volume of ACN significantly enhanced the extraction efficiency of lipophilic pesticides. Similar to the findings of Shin et al. (2020) [19], who demonstrated that a solvent-to-sample ratio of 3:1 or greater substantially improved the recovery of lipophilic pesticides in mealworms, this study found that increasing the volume of acetonitrile led to a marked increase in the number of detectable analytes, especially in small sample sizes.

The extraction of lipophilic (fat-loving) pesticides was effectively enhanced by using a greater volume of acetonitrile (ACN) in this study. This observation aligns with the findings of Lee et al. [23], who demonstrated that using a higher solvent-to-sample ratio facilitates more efficient migration of pesticides from complex insect matrices into the organic layer. A larger volume of ACN promotes efficient partitioning, thereby improving analyte transfer, as the adipose layer in the insects acts as a reservoir for lipophilic compounds.

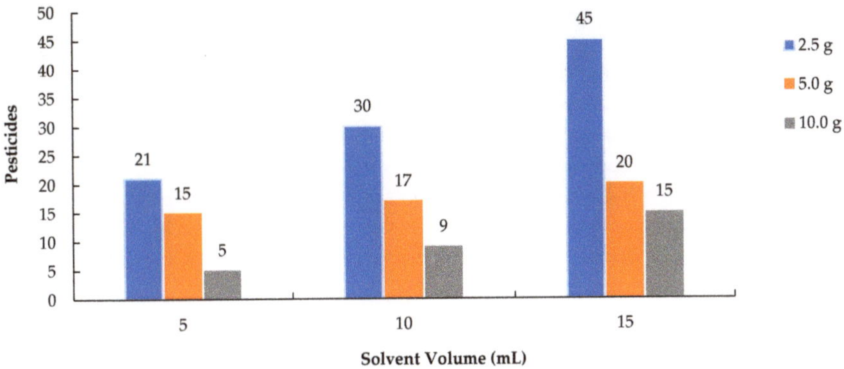

Figure 1. Number of pesticides for extractions with different amounts analyzed by GC-MS/MS.

This study utilized freeze-drying (lyophilization) as a sample preparation technique to enhance analytical precision and maintain the integrity of pesticide residues in edible insect matrices. Freeze-drying efficiently eliminates water content without applying heat, thus reducing the likelihood of thermal degradation or enzymatic modification of target analytes, which is particularly crucial for thermolabile or pH-sensitive pesticides [20]. Moreover, dry samples offer enhanced control over sample weight and solvent ratios, which is crucial for method optimization. Utilizing smaller sample sizes (e.g., 2.5–5.0 g) is especially beneficial for insects, which frequently produce less biomass and possess a higher fat content that may hinder extraction efficiency [19]. Our study confirms that lyophilized insect samples, when adequately rehydrated, provide consistent analyte recoveries and facilitate accurate control over sample-to-solvent ratios [17]. In this study, 5 mL of water was added to all extraction mixtures to hydrate the freeze-dried samples, demonstrating that the addition of water improves the partitioning efficiency of pesticides during QuEChERS extraction by swelling the matrix and enhancing analyte desorption [20,24]. Similarly, studies on food matrices with a low moisture content showed that insufficient hydration can lead to poor recovery due to the inadequate extraction of target compounds trapped within the matrix [25]. The consistent recovery of most pesticides (70–120%) indicates that the water proportion used was sufficient to optimize extraction without excessively diluting the solvent system. However, slight variations in recovery were observed for some lipophilic pesticides, which could be influenced by the balance between solvent volume and water content. Excessive water may reduce the extraction efficiency of non-polar pesticides by increasing the polarity of the extraction environment, thus hindering their transfer into the organic phase.

The quality of the extract was also enhanced by the addition of $MgSO_4$ and Na_3 Citrate, which facilitated phase separation and effectively removed residual water. This improvement is essential when dealing with insect matrices that are high in fat and susceptible to emulsification. These findings are consistent with recent food safety studies that employed QuEChERS-based extraction protocols [10,26].

2.2. *Optimization of MRM Condition for GC–MS/MS Analysis*

To optimize pesticide detection, collision energy (CE) tests were conducted to identify the two most suitable ion transitions—primary and secondary—from precursor to product ions for multiple reaction monitoring (MRM). Quantification was performed by GC-MS/MS using the external standard method. The Agilent Mass Hunter software (version 10.2) was used to determine the peak area of each analyte's primary ion transition.

Precursor ions were identified through full-scan analysis for each pesticide, and their retention times (RTs) were determined under the specified gas chromatography (GC) conditions. Product ion scans were then carried out at varying collision energies (0–50 eV) using the selected precursor ions. The optimal MRM conditions were established by evaluating the sensitivity of the resulting product ions at varying CE levels. For confirmation, two product ions per precursor, along with their ion ratios and retention times, were selected. The qualitative and quantitative ions were selected based on their signal intensity and minimal interference from surrounding ions or baseline noise, as detailed in Table 1, and Figure 2 shows the mass spectra of the representative pesticide standards obtained under collision-induced dissociation (CID) conditions, including the ion ratios for dichlorvos and β-BHC.

Table 1. The MRM Conditions of the primary and secondary transitions of a precursor to product ions for pesticides.

No.	Pesticides	Groups	RT (min)	Precursor > Product Ion (CE, eV)		Dwell
				Quantifier Ion	Qualifier Ion	
1	Methamidophos	OP	4.581	141 > 95 (5 eV)	141> 80 (20 eV)	10
2	Dichlorvos	OP	4.610	184.9 > 93 (10 eV)	144.9 > 109 (10 eV)	10
3	Mevinphos	OP	5.557	127 > 10 (10 eV)	192> 126 (10 eV)	10
4	Acephate	OP	5.663	136 > 94 (15 eV)	78.9 > 47 (10 eV)	10
5	Omethoate	OP	6.850	110 > 79 (15 eV)	156 > 79 (25 eV)	10
6	Fenobucarb	CB	6.769	121 > 77 (20 eV)	149.9 > 121.1 (5 eV)	10
7	Monocrotophos	OP	7.380	127.1 > 109 (10 eV)	127.1 > 95 (15 eV)	10
8	Dimethoate	OP	7.806	87 > 46 (20 eV)	125 > 47 (15 eV)	10
9	Carbofuran	CB	7.781	164.2 > 149.1 (10 eV)	149.1 > 121.1 (5 eV)	10
10	Atrazine	Herbicide	7.847	214.9 > 58.1 (10 eV)	200 > 122.1 (5 eV)	10
11	BHC-beta	OC	7.991	216.9 > 181.1 (5 eV)	218.9 > 183.1 (5 eV)	10
12	Diazinon	OP	8.239	137.1 > 84 (10 eV)	199.1 > 93 (15 eV)	10
13	Pirimicarb	CB	8.685	238 > 166.2 (10 eV)	166 > 55.1 (20 eV)	10
14	Parathion-methyl	OP	9.091	262.9 > 109 (10 eV)	125 > 47 (10 eV)	10
15	Fenitrothion	OP	9.094	125.1 > 47 (15 eV)	277 > 109 (5 eV)	10
16	Pirimiphos-methyl	OP	9.096	232.9 > 151 (5 eV)	232.9 > 125 (5 eV)	10
17	Carbaryl	CM	9.250	144.1 > 116.1 (15 eV)	144.1 > 89 (45 eV)	10
18	Heptachlor	OC	9.279	271.7 > 236.9 (15 eV)	273.7 > 238.9 (15 eV)	10
19	Malathion	OP	9.672	126.9 > 99 (5 eV)	172.9 > 99 (15 eV)	10
20	Aldrin	OC	9.880	262.9 > 192.9 (35 eV)	263 > 228 (20 eV)	10
21	Chlorpyrifos	OP	9.904	196.9 > 169 (15 eV)	198.9 > 171 (15 eV)	10
22	Methidathion	OP	10.952	144.9 > 85 (5 eV)	144.9 > 58.1 (15 eV)	10
23	DDE-o,p′	OC	11.024	246 > 176.2 (30 eV)	248 > 176.2 (30 eV)	10
24	Endosulfan	OC	11.197	158.9 > 89 (35 eV)	194.9 > 124.9 (30 eV)	10
25	Prothiofos	OP	11.447	113 > 94.9 (10 eV)	266.9 > 239 (5 eV)	10
26	Profenofos	OP	11.500	207.9 > 63 (30 eV)	338.8 > 268.7 (15 eV)	10
27	DDE-p,p′	OC	11.569	246.1 > 176.2 (30 eV)	246 > 211 (15 eV)	10
28	DDT-o,p′	OC	12.382	235 > 165.2 (20 eV)	235 > 199.1 (15 eV)	10
29	Ethion	OP	12.382	231 > 129 (10 eV)	231 > 175 (5 eV)	10
30	Triazophos	OP	12.609	161.2 > 134.2 (5 eV)	161.2 > 91 (15 eV)	10
31	Triphenyphosphate	OP	13.315	214.9 > 168.1 (15 eV)	232.9 > 215.1 (10 eV)	10
32	Carbosulfan	CM	13.778	118 > 76 (5 eV)	164 > 103.1 (25 eV)	10
33	EPN	OP	13.897	169 > 141.1 (5 eV)	169 > 77.1 (25 eV)	10
34	Phosmet	OP	13.959	160 > 77.1 (20 eV)	160 > 133.1 (10 eV)	10
35	Amitraz	Acaricide/Insecticide	14.714	132.1 > 117.1 (15 eV)	162 > 132.2 (5 eV)	10
36	Azinphos-methyl	OP	14.610	160 > 77 (20 eV)	132.1 > 77 (15 eV)	10
37	Cyhalothrin (Lambda)	PY	14.834	208.1 > 181.1 (10 eV)	181.1 > 152.1 (30 eV)	10

Table 1. *Cont.*

No.	Pesticides	Groups	RT (min)	Precursor > Product Ion (CE, eV)		Dwell
				Quantifier Ion	Qualifier Ion	
38	Fenpropathrin	PY	14.834	181.1 > 152.1 (25 eV)	181 > 127 (5 eV)	10
39	Azinphos-ethyl	OP	15.238	132 > 77.1 (15 eV)	160 > 77.1 (20 eV)	10
40	Permethrin	PY	15.688	183.1 > 168.1 (10 eV)	183.1 > 153.1 (15 eV)	10
41	Coumaphos	PY	15.844	210 > 182 (10 eV)	361.9 > 109 (15 eV)	10
42	Cyfluthrin I	PY	16.172	162.9 > 90.9 (15 eV)	226 > 206 (5 eV)	10
43	Cyfluthrin II	PY	16.253	162.9 > 90.9 (15 eV)	162.9 > 127 (5 eV)	10
44	Cyfluthrin III	PY	16.254	162.9 > 90.9 (15 eV)	162.9 > 127 (5 eV)	10
45	Cyfluthrin IV	PY	16.367	162.9 > 90.9 (15 eV)	162.9 > 127 (5 eV)	10
46	Cypermethrin I	PY	16.454	163 > 91 (10 eV)	226 > 206 (5 eV)	10
47	Cypermethrin II	PY	16.543	163.1 > 127.1 (5 eV)	163.1 > 91 (15 eV)	10
48	Cypermethrin III	PY	16.628	163.1 > 127.1 (5 eV)	163.1 > 91 (15 eV)	10
49	Cypermethrin IV	PY	16.651	163.1 > 127.1 (5 eV)	163.1 > 91 (15 eV)	10
50	Fenvalerate	PY	17.561	167 > 125.1 (5 eV)	208.9 > 141.1 (15 eV)	10
51	Fluvalinate-tau	PY	17.558	250 > 55 (40 eV)	181 > 152 (40 eV)	10
52	Esfenvalerate	PY	17.560	167 > 125.1 (10 eV)	181 > 152.1 (25 eV)	10
53	Deltamethrin	PY	18.082	252.9 > 93 (15 eV)	181 > 152.1 (25 eV)	10

RT = retention time, CE = collision energy, eV = electron volt. Pesticide groups: organophosphates (OPs), carbamates (CBs), pyrethroids (PYs), organochlorines (OCs).

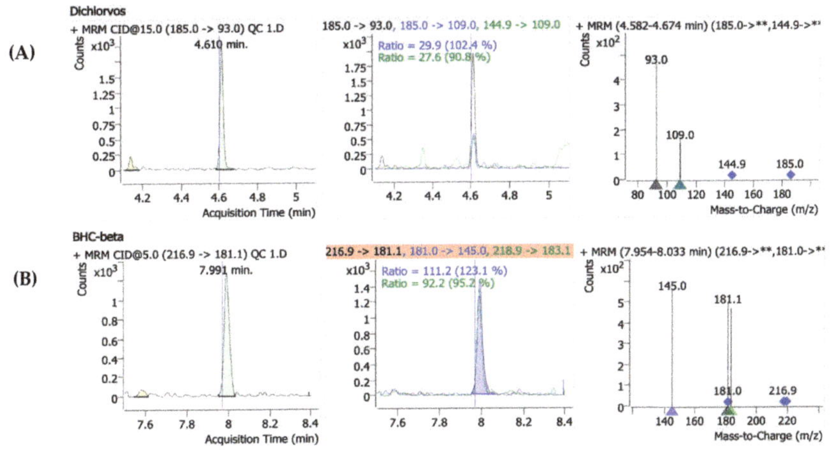

Figure 2. The representative mass spectra of the selected pesticide standards obtained under collision-induced dissociation (CID) conditions. (**A**) Dichlorvos** The protonated molecular ion $[M+H]^+$ at m/z 185.0 yields fragment ions at m/z 93.0 and 109.0. (**B**) BHC-beta** The precursor ion $[M+H]^+$ at m/z 216.9 generates a fragment ion at m/z 181.1, while another ion at m/z 218.9 fragments into m/z 183.1.

2.3. *Method Validations*

2.3.1. Linearity, Selectivity, Limit of Detection, and Limit of Quantification

The selectivity of the analytical method was evaluated by comparing chromatograms of the recovery samples with those of standard solutions and blank samples. All 47 pesticide compounds demonstrated consistent retention times and m/z values under the established method using matrix-matched standards, with no interfering peaks observed. These results confirm the high selectivity and effective separation achieved by the developed analytical technique. The limits of detection (LODs) for individual pesticides ranged from 1 to 5 μg/kg, while the limits of quantification (LOQs) ranged from 10 to 15 μg/kg. Linearity, which refers to the method's ability to produce results directly proportional to the analyte concentration within a specified range, was assessed using matrix-matched calibration curves. The GC–MS/MS results, summarized in Tables 2 and 3, include correlation coefficients (R^2), y-intercepts, and calibration curve parameters. The R^2 values ranged from 0.9940 to 0.9999, and the linearity of the method was evaluated by assessing the percentage deviation of the back-calculated concentration (BBC) from the nominal concentration at each calibration level. The % deviations of the BBC at each calibration level were calculated and reported following the criteria specified in Table 2, which were used according to the SANTE guidelines [24]. The results demonstrated that the majority of the pesticide compounds (47 analytes) had acceptable limits across all levels, indicating excellent linearity. Most compounds exhibited deviation values below ±30%, while, at mid- to high levels (10–500 µg/kg), deviations generally fell within ±20%. These findings confirm that the analytical method demonstrates excellent linearity for quantitative analysis.

Table 2. The linearity of the analytical method based on the % deviation of back-calculated concentrations (BBCs) in edible insect samples (*n* = 5).

No.	Pesticides	Calibration Levels (µg/kg)	% Deviation of BBC						
			5 µg/kg	10 µg/kg	25 µg/kg	50 µg/kg	100 µg/kg	250 µg/kg	500 µg/kg
1	Methamidophos	5–500	15.00	5.90	2.12	−15.87	−13.28	4.23	−0.45
2	Dichlorvos	5–500	7.60	11.11	−1.91	−3.21	−3.61	−4.44	1.26
3	Acephate	5–500	−13.20	16.59	−5.28	5.86	−0.79	−12.56	0.49
4	Mevinphos	5–500	4.00	2.30	−6.80	−2.06	−6.15	−2.95	−3.15
5	Fenobucarb	5–500	8.60	1.65	11.90	−8.64	−3.81	2.94	1.44
6	Omethoate	5–500	10.80	12.00	−14.03	−11.17	−5.59	−1.67	0.47
7	Monocrotophos	5–500	5.40	10.88	−13.48	−3.93	−8.83	4.23	0.50
8	Dicrotofos	5–500	−14.60	18.62	−13.35	4.09	−8.22	−4.05	0.51
9	Carbofuran	5–500	14.91	−8.25	−0.65	8.79	8.27	−1.18	−0.09
10	Dimethoate	5–500	8.60	−5.59	−22.40	−1.31	2.02	−1.13	0.40
11	Atrazine	5–500	−2.27	17.08	4.72	−4.64	−2.36	1.03	−0.14
12	BHC-beta	5–500	−15.20	−26.36	9.23	−9.91	−15.13	−1.19	0.93
13	Diazinon	5–500	3.20	−15.44	−8.74	−1.99	4.02	4.06	−1.12
14	Pirimicarb	5–500	14.14	−4.21	−6.57	−2.36	2.40	3.50	−0.92
15	Parathion-methyl	5–500	4.07	−6.45	−10.91	2.94	1.70	5.34	0.34
16	Fenitrothion	5–500	4.40	−9.73	−18.16	−11.32	4.45	5.69	0.31
17	Pirimiphos-methyl	5–500	−1.20	−9.95	9.16	−10.72	9.20	−4.14	0.42
18	Carbaryl	5–500	13.60	−15.57	−13.24	−12.33	−5.16	−0.59	0.50
19	Heptachlor	5–500	−13.00	−16.18	1.72	−4.09	−7.12	−0.28	2.34
20	Malathion	5–500	−11.20	0.69	−8.04	−11.93	−6.05	−1.79	0.29
21	Aldrin	5–500	5.00	1.35	−11.80	−1.96	6.12	6.94	−1.91
22	Chlorpyrifos	5–500	9.48	11.20	1.05	−7.10	−6.10	0.28	0.22
23	Methidathion	5–500	4.40	16.46	−13.08	−4.76	7.63	−1.86	0.41
24	DDE-o,p′	5–500	2.40	6.28	−11.70	0.48	8.06	6.38	−1.86
25	Endosulfan	5–500	15.60	3.30	7.55	16.86	18.97	14.97	−0.86
26	Prothiofos	5–500	0.60	17.41	4.05	−10.45	−6.54	−0.09	0.35
27	Profenofos	5–500	−0.40	13.88	11.71	−16.18	4.99	−15.45	1.06
28	DDE-p,p′	5–500	−17.80	−9.33	−13.26	−0.98	5.47	6.24	−1.71
29	DDT-o,p′	5–500	−17.80	−19.65	18.98	0.70	−16.90	−5.07	1.84
30	Ethion	5–500	15.20	19.46	18.70	−13.01	−0.40	−6.08	2.34
31	Triazophos	5–500	12.22	−19.42	−9.26	−0.41	−7.67	−5.62	0.37
32	Cabosulfan	5–500	2.21	−6.79	15.80	−11.43	−2.20	−17.52	0.91
33	EPN	5–500	−11.44	3.45	−14.72	−11.45	−1.77	−7.85	2.81
34	Phosmet	5–500	−20.90	12.31	11.61	5.14	−5.51	7.81	0.47
35	Amitraz	5–500	15.93	1.58	1.09	−4.15	−6.42	−0.08	0.29
36	Azinphos-methyl	5–500	−2.00	19.70	−16.77	−7.35	−14.98	−2.13	−2.76
37	Cyhalothrin (Lambda)	5–500	4.22	1.87	18.73	−10.88	−16.34	−5.26	1.97
38	Fenpropatrin	5–500	−12.39	−0.15	17.54	−10.73	−17.61	−3.45	1.57
39	Azinphos-ethyl	5–500	1.39	−12.96	−0.72	−9.53	−3.62	0.66	0.43
40	Permethrin	5–500	−11.92	10.89	−0.34	−5.97	−4.53	2.37	−0.36
41	Coumaphos	5–500	7.29	6.81	−12.20	5.00	3.23	−1.04	0.50
42	Cyfluthrin	5–500	−13.02	0.17	5.09	−6.12	−17.69	−0.59	0.69

Table 2. *Cont.*

No.	Pesticides	Calibration Levels (μg/kg)	% Deviation of BBC						
			5 μg/kg	10 μg/kg	25 μg/kg	50 μg/kg	100 μg/kg	250 μg/kg	500 μg/kg
43	Cypermethrin	5–500	−6.64	5.29	3.66	−3.77	3.16	−6.58	1.64
44	Fenvalerate	5–500	2.93	1.41	8.03	−14.12	3.37	−6.92	1.81
45	Esfenvalerate	5–500	12.69	5.08	−3.55	−15.66	−14.34	−5.73	2.05
46	Fluvalinate-tau	5–500	2.55	14.84	4.57	−11.90	0.08	3.79	−0.67
47	Deltamethrin	5–500	−14.03	5.22	−6.69	−17.73	−11.12	−1.41	0.87

BBC = back-calculated concentration. % deviation of the back-calculated concentration from the true concentration ≤ ±20.

Table 3. Limit of detection (LOD), limit of quantification (LOQ), recovery, precision, and matrix effect in pooled edible insect samples by GC-MS/MS (n = 5).

No.	Pesticides	Linearity (r^2)	Limit of Detection Limit (LOD)	Limit of Quantification (LOQ)	Low (10 μg/kg) Spike Level		Medium (100 μg/kg) Spike Level		High (500 μg/kg) Spike Level		ME %
			μg/kg	μg/kg	Recovery (%)	RSD (%)	Recovery (%)	RSD (%)	Recovery (%)	RSD (%)	
1	Methamidophos	0.9979	1	10	86.72	3.77	122.12	1.96	99.55	2.81	6.13
2	Dichlorvos	0.9989	1	10	96.39	2.71	98.09	2.15	101.26	2.24	−1.60
3	Acephate	0.9939	2	10	64.54	4.77	94.72	3.32	100.49	3.40	14.33
4	Mevinphos	0.9948	1	10	100.37	2.57	87.57	2.39	98.53	2.23	−15.40
5	Fenobucarb	0.9986	2	10	96.19	2.63	111.90	2.08	101.44	2.25	−33.00
6	Omethoate	0.9936	3	15	88.50	4.42	85.97	3.72	100.47	2.99	24.24
7	Monocrotophos	0.9938	2	10	80.12	3.85	86.52	4.62	100.5	2.98	−9.59
8	Dicrotofos	0.9934	5	10	87.40	2.50	86.65	3.14	100.51	3.04	3.13
9	Carbofuran	0.9993	1	10	108.27	3.82	99.35	2.61	99.91	2.38	−11.80
10	Dimethoate	0.9958	1	10	70.38	4.33	82.73	2.48	100.4	2.39	−2.18
11	Atrazine	0.9999	1	10	97.64	3.18	104.72	1.86	99.86	2.29	−13.40
12	BHC-beta	0.9979	1	10	84.87	2.38	109.23	2.06	100.93	2.24	−4.21
13	Diazinon	0.9991	2	10	104.02	2.47	91.26	1.90	98.88	2.16	9.90
14	Pirimicarb	0.9994	1	10	102.40	2.88	93.43	2.28	99.08	2.22	0.22
15	Parathion-methyl	0.9970	1	10	86.40	2.57	89.09	2.32	100.34	2.89	−17.40
16	Fenitrothion	0.9975	3	10	85.00	3.23	81.84	2.43	100.31	3.07	6.06
17	Pirimiphos-methyl	0.9953	1	10	79.40	4.52	109.16	3.18	100.42	2.74	−10.08
18	Carbaryl	0.9969	1	10	94.84	2.83	86.76	2.03	100.5	2.13	8.68
19	Heptachlor	0.9961	1	10	92.88	4.49	101.72	4.64	102.34	3.10	4.74
20	Malathion	0.9978	1	10	100.00	2.45	91.96	2.12	100.29	2.25	4.63
21	Aldrin	0.9975	3	10	106.12	4.16	88.20	2.04	98.09	2.14	−3.57
22	Chlorpyrifos	0.9996	3	10	93.90	2.69	101.05	1.96	100.22	2.21	−5.01
23	Methidathion	0.9957	1	10	72.25	2.54	86.92	2.37	100.41	2.36	−4.82
24	DDE-o,p′	0.9975	1	10	108.06	3.11	88.30	1.94	98.14	2.19	6.32
25	Endosulfan	0.9957	3	10	118.97	5.14	107.55	3.48	99.14	4.95	23.70
26	Prothiofos	0.9994	5	10	93.46	2.33	104.05	1.94	100.35	2.17	−33.01
27	Profenofos	0.9944	5	10	74.99	2.89	111.71	2.28	101.06	2.31	1.02
28	DDE-p,p′	0.9980	1	10	105.47	2.72	86.74	2.04	98.29	2.29	−1.30
29	DDT-o,p′	0.9967	1	10	83.10	2.61	118.98	1.97	101.84	2.14	2.16
30	Ethion	0.9941	1	10	79.60	2.33	118.70	2.07	102.34	2.11	10.49
31	Triazophos	0.9967	2	10	83.45	2.49	90.74	2.40	100.37	2.32	3.07
32	Cabosulfan	0.9959	3	10	77.80	2.36	115.80	2.37	100.91	2.48	−2.53
33	EPN	0.9926	1	10	78.23	3.52	98.62	3.74	102.81	4.02	3.45
34	Phosmet	0.9968	3	10	64.54	2.88	111.61	2.43	100.47	2.38	−5.33
35	Amitraz	0.9996	1	10	93.58	2.45	101.09	1.95	100.29	2.17	5.85
36	Azinphos-methyl	0.9909	3	10	85.02	2.83	83.23	2.41	97.24	2.71	19.98
37	Cyhalothrin (Lambda)	0.9960	3	10	83.66	2.67	118.73	1.82	101.97	2.14	−0.88
38	Fenpropatrin	0.9966	3	10	82.39	2.62	117.54	2.03	101.57	2.21	3.62
39	Azinphos-ethyl	0.9954	1	10	80.30	2.82	99.28	2.43	100.43	2.39	9.48
40	Permethrin	0.9996	5	10	95.47	2.87	99.66	2.21	99.64	2.06	14.19
41	Coumaphos	0.9940	1	10	89.40	2.53	87.80	2.50	100.5	2.25	−20.58
42	Cyfluthrin	0.9961	2	10	92.01	3.61	106.86	2.97	101.56	6.02	−0.40
43	Cypermethrin	0.9976	2	10	84.20	4.08	109.24	2.62	101.17	2.45	15.64
44	Fenvalerate	0.9967	1	10	77.04	2.63	109.37	1.91	101.76	2.25	−16.40
45	Esfenvalerate	0.9959	2	10	85.66	2.54	120.45	1.97	102.05	2.33	3.56
46	Fluvalinate-tau	0.9982	3	10	100.08	2.71	104.57	1.86	99.33	2.38	−3.01
47	Deltamethrin	0.9982	5	10	88.88	2.52	98.63	2.11	100.87	3.84	0.48

RSD = relative standard deviation. ME = matrix effect was classified as: suppression (%ME < −20%), enhancement (%ME > +20%), and no effect (%ME between −20% and +20%).

2.3.2. Precision and Accuracy

The accuracy and precision of the target compounds in the developed method were assessed based on average recovery and relative standard deviation (RSD) from five trials

($n = 5$). Three spiking concentrations—low, medium, and high—were selected based on the linear range of the target compounds. For a linear range of 5 to 500 µg/kg, the spiking levels were set at 10 µg/kg (low), 100 µg/kg (medium), and 500 µg/kg (high). Recovery rates ranged from 70% to 120%, with RSD values below 20%, indicating the method's effectiveness. In the recovery test using GC–MS/MS for edible insects, all pesticides exhibited recovery within the acceptable range of 70–120%. Additionally, RSD values were confirmed to be under 20% for all pesticides, as shown in Table 3.

In our study, acetonitrile was selected as the extraction solvent for the multiresidue analysis of pesticides in edible insect samples. In addition, to achieve better efficiency in the extraction and recovery of multiresidue pesticides from the samples, the QuEChERS procedure was employed in our study, as all recovery rates fell within the range considered effective and promising. The QuEChERS procedure yielded extraction recoveries ranging from 64.54% to 122.12%, with RSDs between 1.86% and 6.02%. On the other hand, the QuEChERS procedure may have caused strong salting-out effects, as the mean recoveries of methamidophos (122.12%) treated at a level of 100 µg/kg were observed to be higher than those of other pesticides [19,27]. However, such salting-out effects were not observed in our study. Therefore, the recoveries of multiresidue pesticides were generally between 70 and 120% with RSDs below 20%, indicating that the proposed method in this study was feasible to analyze 47 pesticides in samples.

The validated method developed in this study demonstrates performance metrics comparable to and, in some aspects, exceeding those reported in the recent literature. The technique achieved strong linearity (R^2 = 0.9940–0.9999) across all 47 target pesticides, which aligns closely with the other studies in Table 4, and Shin et al. (2020) [19] and Labu et al. (2022) [6], both of which reported R^2 values > 0.995. The LOQ range (10–15 µg/kg) in this study is slightly higher than the 5 µg/kg commonly reported in several other investigations, such as those by Kim et al. (2020) [17] and Kolakowski et al. (2021) [7], yet remains within acceptable sensitivity limits for regulatory purposes. In terms of recovery, this study showed results from 64.54% to 122.12%, with over 97.87% of the analytes falling within the 70–120% range defined by SANTE. And, relative standard deviations (RSDs) in this study were consistently below 6.02%, indicating excellent repeatability.

Moreover, the relative response of each analyte varied according to its molecular structure and fragmentation pattern. Nevertheless, the sensitivity of the developed GC-MS/MS method was sufficient to screen GC-amenable pesticides at a fortification level of 10 µg/kg using a 2.5 g sample and 15 mL of solvent for extraction. Minimal matrix interference was observed under these conditions, as demonstrated by the comparison of total ion chromatograms between blank edible insect samples and those fortified at 100 and 500 µg/kg (Figure 3). These results confirm the method's suitability for analyzing trace levels of pesticides in complex, high-fat matrices.

2.3.3. Matrix Effect

The matrix effect provided information about the matrix effects. Depending on the decrease/increase in the percentage of the slope, different matrix effects could be observed: a signal suppression or enhancement effect between −20% and 0% and between 0% and +20% was considered to be mild; a medium effect was supposed when the slope values were between −50% and −20% or +20% and +50%; and a strong effect of signal suppression or enhancement was supposed when values were below −50% or above +50%. These distributions depended not only on the matrix effect but also on the combination of the compound matrix. Pesticides with %ME values within the acceptable range of −20% to +20% experience minimal matrix effects and are less affected by matrix interferences, ensuring more reliable quantification [27–29].

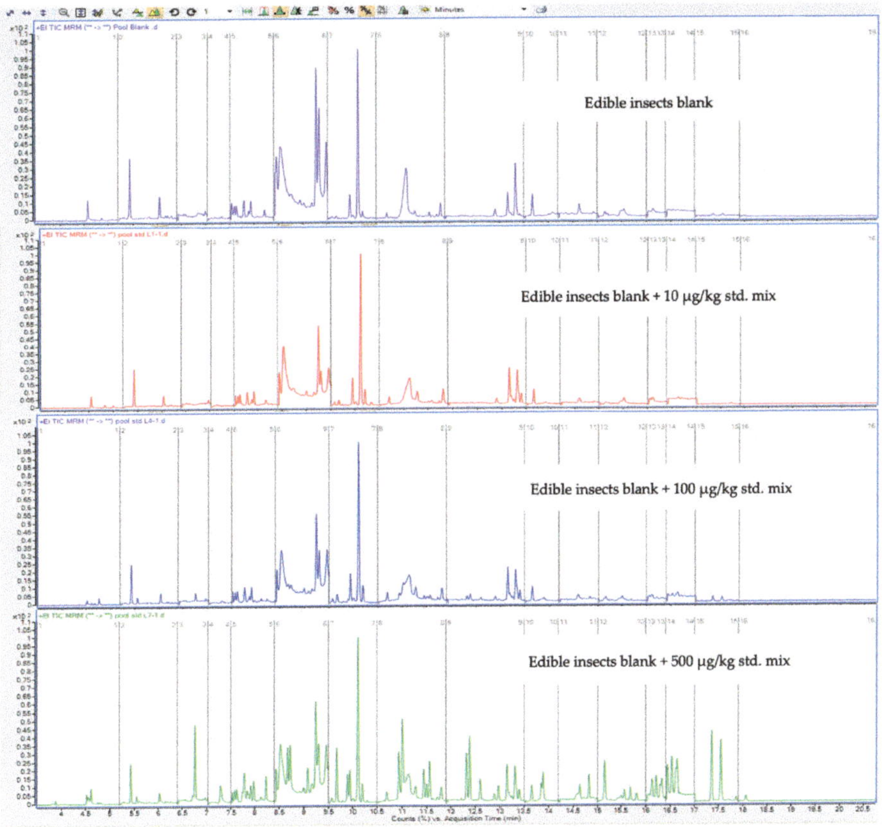

Figure 3. Reconstructed GC-MS/MS chromatograms of edible insects blank and edible insects blank fortified at 10, 100, and 500 µg/kg in 20 min analyses of 47 target compounds with 1 µL injection volume.

Matrix effects (%MEs) observed in the analysis of pesticide residues in edible insect samples varied widely among the compounds, ranging from −33.01% to 24.04%. For instance, fenobucarb and prothiofos exhibited high matrix suppression effects at −33.00% and −33.01%. While omethoate was at +24.24% and endosulfan at +23.70%, respectively, more than 95.74% of the pesticides showed a soft matrix effect (Table 2) with negligible effects in the tested range [29]. Moreover, our findings are in agreement with the study by Łozowicka et al. (2017), who emphasized that modified QuEChERS protocols incorporating buffering salts and increased solvent volumes effectively reduce matrix effects during GC-MS/MS analysis in complex matrices such as soil or insects [10]. However, most of the pesticides in the current investigation were unaffected by the matrix, likely due to the edible insect matrices being effectively eliminated. The extraction procedure effectively removed numerous proteins and lipids that significantly impaired the matrix effect. The dilution procedure used to prepare the sample can also be helpful. Between the extraction and partitioning stages, 5–0 times more solvent was utilized than in traditional QuEChERS procedures [10,30]. Dilution reduced the concentration of the sample matrices to a point where the signal remained unaffected. However, using this method, a small percentage of pesticides (4.25%) exhibited a moderate matrix effect (Table 3). Therefore, for accurate quantification, a matrix-matched calibration procedure ought to be employed.

Table 4. Previous studies' analytical methods for determining pesticide residues in edible insects.

Matrix	Instrument	No. Analysis	RSD (%)	Recovery (%)	LOQ (µg/kg)	Linearity (r^2)	Matrix Effect (ME %)	Reference
Mealworms	LC-MS/MS	353	5–15	75–115	5	>0.995	−20 to +25	[19]
Mealworms	GC-MS/MS, LC-MS/MS	300	5–12	80–115	5	>0.993	−15 to +20	[17]
Edible insects	GC-MS/MS, LC-MS/MS	-	5–20	75–120	5	>0.990	−18 to +22	[7]
Edible insects (6 species)	LC-MS/MS	374	6–15	75–110	5	>0.995	−15 to +25	[6]
Mealworm larvae	GC-MS/MS	247	0–19.9	70–120	50	≥0.990	Not specified	[31]

3. Materials and Methods

3.1. Chemicals and Materials

Standard pesticide mixtures with a purity of at least 99% were obtained from Dr. Ehrenstorfer (Augsburg, Germany). The combinations comprised a total of 47 chemicals. A diverse array of chemical classes was incorporated—organophosphates (OPs), carbamates (CBs), pyrethroids (PYs), and organochlorines (OCs). Pesticides commonly used in agricultural activities particularly relevant to crops and associated with edible insect habitats were prepared at a concentration of 1000 µg/mL in acetonitrile. For standard fortification, a composite pesticide stock solution was established with various concentrations of standard curve pesticides ranging from 5 to 500 µg/L. A standard mix was used to fortify 2.5 g of pooled edible insect blank by adding 100 µL of the mix to the sample prior to extraction. HPLC-grade acetonitrile, ethyl acetate, and water were obtained from J.T. Baker (Avantor, Radnor Township, PA, USA). They were used for GC/MS and solvent extraction. QuEChERS original packet 50 mL centrifuge tubes filled with 4 g of $MgSO_4$, 1 g of NaCl, 1 g of $Na_3C_6H_5O_7{\cdot}2H_2O$, and 0.5 g of $Na_2HC_6H_5O_7{\cdot}1.5H_2O$ from Agilent Technologies, Santa Clara, CA, USA. Additionally, dispersive cleanup tubes (2 mL) were sourced, containing 150 mg of $MgSO_4$, 50 mg of PSA sorbent, bulk carbograph, and 50 mg of end-capped C-18 sorbent, also from Agilent Technologies, USA. For GC-MS/MS, helium (99.999%) and nitrogen (99.999%) served as the carrier gas and collision gas, respectively.

3.2. Sample Selection

In June and July 2023, edible insects were gathered from a local market in Chiang Mai, including bamboo caterpillars, house crickets, silkworm pupae, giant water bugs (*Lethocerus indicus*), and grasshoppers. A total of 200 g of edible insect samples were collected and freeze-dried at −50 °C for at least 72 h using an SP VirTis Genesis Pilot Freeze Dryer and the Genesis 25 L Pilot Lyophilizer (Westminster, CO, USA). After freeze-drying, the samples were ground into a fine powder. A sample blank was produced by combining multiple individual insect specimens into a single, homogenized unit in studies that involve consumable insects. The non-fortified sample was also prepared to create a blank matrix for matrix-matched standards. Each sample, weighing 2.5 g, was placed in a 50 mL centrifuge tube (Corning Inc., Wujiang, China) and stored at −20 °C until analysis.

3.3. Sample Extraction

Pesticide residues in the edible insect samples were extracted using the EN QuEChERS method [26], with slight modifications. Prior to extraction, the samples were brought to room temperature and spiked with an appropriate standard mixture to achieve final concentrations of 10, 100, and 500 µg/kg. A non-fortified (blank) sample was also prepared to serve as a blank matrix for matrix-matched standards. For the extraction, 5 mL of purified water and 15 mL of acetonitrile were added to each sample tube and mixed thoroughly.

The tubes were then tightly capped and shaken for 5 min using a SPEX 2000 Geno Grinder (Digital Multi-Tube Vortex Mixer, Model VXMTDG, OHAUS, Parsippany, Morris County, NJ, USA) at 2500 rpm. Following this, a QuEChERS extraction packet containing 4 g of $MgSO_4$, 1 g of NaCl, 1 g of $Na_3C_6H_5O_7{\cdot}2H_2O$, and 0.5 g of $Na_2HC_6H_5O_7{\cdot}1.5H_2O$ was added. The contents were shaken again for 5 min at the same speed and then centrifuged at 2000 rpm for 5 min.

Next, approximately 1 mL of the upper (organic) layer of the extract was transferred into a 2 mL dispersive solid-phase extraction (SPE) tube containing 150 mg of $MgSO_4$, 50 mg of PSA sorbent, bulk carbograph, and 50 mg of end-capped C18 sorbent. The tube was capped, vortexed for 1 min, and centrifuged at 1300 rpm for 3 min. The resulting upper layer was filtered through a 0.45 µm syringe filter and transferred to sample vials for analysis. Finally, 1 µL was injected into a Gas Chromatography Triple Quadrupole Mass Spectrometer (GC-MS/MS) using an autosampler.

3.4. GC-MS/MS Instrument Conditions

GC-MS/MS analysis was performed using an Agilent 8890 Gas Chromatograph, equipped with a 7693A autosampler and a 7000E triple quadrupole mass spectrometer, operated through Mass Hunter Software (Qualitative Analysis version B.12.0.430.0 and Quantitative Analysis version B.12.1.938.3) installed on a dedicated workstation for data acquisition and processing (Agilent Technologies, Inc., Santa Clara, CA, USA). The GC and backflush method parameters, including the tandem mass spectrometry and MRM (multiple reaction monitoring) mode settings, are detailed in Tables 5 and 6. Analyte separation was achieved using two HP-5 ms Ultra Inert capillary columns from Agilent (0.25 mm internal diameter × 30 m length, 0.25 µm film thickness), connected via a backflush union.

Table 5. The GC and backflush method conditions.

Parameter	Condition		
Inlet	MMI Injection mode Spitless		
Carrier gas	Helium		
Inlet flow	1 mL/min		
Inlet temperature	280 °C		
Injection volume	1 µL		
MS transfer line temperature	280 °C		
Oven program	Ramp Time	Temp	Hold
	60 °C	1 min	40 °C/min
	170 °C	0 min	10 °C/min
	310 °C	3 min	
Total Run Time	20.75 min		

Table 6. The tandem mass spectrometer and MRM mode condition.

Parameter	Condition
Mode	Electron impact
Transfer line temperature	280 °C
Source temperature	280 °C
Quadrupole temperature	Q1 and Q2 = 150 °C
MS1 resolution	Wind
MS2 resolution	Wind
Collision gas flow	Nitrogen at 1.5 mL/min
Quenching gas flow	Helium at 2.25 mL/min
Detector Gain	10

3.5. Method Validation

3.5.1. Linearity and Sensitivity

The linearity of all compounds in both the solvent and mixed standard solutions was evaluated by plotting the peak area against standard concentrations of 5, 10, 25, 50, 100, 250, and 500 µg/kg. Calibration curves were constructed using a 1/x weighting factor, excluding the origin. The coefficient of determination (R^2) was calculated to assess the accuracy of the calibration model, with an acceptance criterion of $R^2 \geq 0.9900$. Moreover, the linearity of the method was evaluated by assessing the percentage deviation of the back-calculated concentration (BBC) from the nominal concentration at each calibration level. According to the criteria, the deviation should not exceed ±20% at all concentration levels. Instrument sensitivity was assessed by determining the limit of detection (LOD) and the limit of quantification (LOQ). The LOD was defined as the lowest concentration at which a pesticide signal could be reliably distinguished from the background noise at the corresponding retention time. Following the SANTE/11312/2021 guidelines [24], signal-to-noise (S/N) ratios of 3 and 10 were used to define the LOD and LOQ, respectively.

3.5.2. Precision and Accuracy

The accuracy of the method was evaluated through recovery studies, while precision was assessed based on the relative standard deviation (RSD) [24]. For this purpose, a mixture of standard solutions was spiked into a pooled blank sample of edible insects. Recovery and precision were determined at three fortification levels—10, 100, and 500 µg/kg ($n = 3$ for each level)—using matrix-matched standards. The precision of the method was further confirmed by calculating RSD of the analyte responses across replicates.

3.5.3. Matrix Effect

The matrix effect (ME) of each analyte was assessed using a mass spectrometry-based analytical method [24]. To determine the ME of the pesticides in each sample, the linear regression slope of the matrix-matched calibration curve was compared to the linear regression slope of the standard solution calibration without the matrix using the following Equation (1).

$$\text{Matrix effect } \% = \left[\frac{\text{Slope of matrix matched calibration curve}}{\text{Slope of standard solution calibration curve}}\right] - 1 \times 100 \qquad (1)$$

4. Conclusions

A thorough multiresidue analytical technique was devised and validated using GC-MS/MS for the simultaneous identification of 47 pesticides in high-fat edible insect matrices. The refined QuEChERS extraction technique, utilizing acetonitrile and dispersive solid-phase extraction with $MgSO_4$, PSA, and C18, facilitated the efficient elimination of lipids and moisture while maintaining analyte integrity. Method validation was performed in accordance with SANTE guidelines. Most pesticides exhibited adequate recovery rates (70–120%) and relative standard deviations of less than 5% overall. Despite several chemicals demonstrating reduced recovery at the lowest concentration level, the overall technique performance stayed within acceptable regulatory parameters. Matrix-matched calibration proficiently mitigated matrix effects, guaranteeing precise quantification. This established technology is suitable for regular pesticide monitoring in consumable insects, supporting food safety oversight and regulatory compliance.

Author Contributions: Conceptualization, S.H.; methodology, A.W. and N.S.; validation, P.T., K.S., S.K. and U.J.; formal analysis, P.T.; investigation, C.J., B.C. and S.H.; data curation, B.C. and S.H.; writing—original draft preparation, P.T.; writing review and editing, A.W., B.C. and S.H.; visualization, N.S. and S.H.; supervision, S.H.; project administration, P.T.; funding acquisition, S.H. All authors have read and agreed to the published version of the manuscript.

Funding: This research was funded by the School of Health Sciences Research, Research Institute for Health Sciences, Chiang Mai University, Chiang Mai, Thailand, No. 015/2566, and this research was also supported by National Research Foundation of Korea (NRF 2018R1A6A1A03024862), and a TA/RA Scholarship from Chiang Mai University.

Institutional Review Board Statement: Not applicable.

Informed Consent Statement: Not applicable.

Data Availability Statement: The original contributions presented in this study are included in the article. Further inquiries can be directed to the corresponding author.

Acknowledgments: All laboratory facilities were supported by Research Institute for Health Sciences, Chiang Mai University (RIHES, CMU), and the authors would like to thank Chiang Mai University for facilitating this study.

Conflicts of Interest: The authors declare that they have no competing interests.

References

1. van Huis, A.; Oonincx, D.G.A.B. The environmental sustainability of insects as food and feed. A review. *Agron. Sustain. Dev.* **2017**, *37*, 43. [CrossRef]
2. Puwastien, P.; Attig, G. Edible insects in Thailand: An unconventional protein source? *Ecol. Food Nutr. Ecol. Food Nutr.* **1997**, *36*, 133–149. [CrossRef]
3. Rumpold, B.A.; Schlüter, O.K. Nutritional composition and safety aspects of edible insects. *Mol. Nutr. Food Res.* **2013**, *57*, 802–823. [CrossRef] [PubMed]
4. Klunder, H.; Wolkers-Rooijackers, J.C.M.; Korpela, J.; Nout, M.J. Microbiological aspects of processing and storage of edible insects. *Food Control* **2012**, *26*, 628–631. [CrossRef]
5. Hoang, H. Consumer acceptability of alternative foods: A study of processed cricket-based foods in Vietnam. *IOP Conf. Ser. Earth Environ. Sci.* **2023**, *1155*, 012025. [CrossRef]
6. Labu, S.; Subramanian, S.; Cheseto, X.; Akite, P.; Kasangaki, P.; Chemurot, M.; Tanga, C.M.; Salifu, D.; Egonyu, J.P. Agrochemical Contaminants in Six Species of Edible Insects from Uganda and Kenya. *Curr. Res. Insect Sci.* **2022**, *2*, 100049. [CrossRef]
7. Kolakowski, B.M.; Johaniuk, K.; Zhang, H.; Yamamoto, E. Analysis of Microbiological and Chemical Hazards in Edible Insects Available to Canadian Consumers. *J. Food Prot.* **2021**, *84*, 1575–1581. [CrossRef]
8. Kim, T.K.; Yong, H.I.; Kim, Y.B.; Kim, H.W.; Choi, Y.S. Edible Insects as a Protein Source: A Review of Public Perception, Processing Technology, and Research Trends. *Food Sci. Anim. Resour.* **2019**, *39*, 521–540. [CrossRef]
9. Lee, J.; Kim, L.; Shin, Y.; Lee, J.; Lee, J.; Kim, E.; Moon, J.K.; Kim, J.H. Rapid and Simultaneous Analysis of 360 Pesticides in Brown Rice, Spinach, Orange, and Potato Using Microbore GC-MS/MS. *J. Agric. Food Chem.* **2017**, *65*, 3387–3395. [CrossRef]
10. Anastassiades, M.; Lehotay, S.J.; Štajnbaher, D.; Schenck, F.J. Fast and Easy Multiresidue Method Employing Acetonitrile Extraction/Partitioning and "Dispersive Solid-Phase Extraction" for the Determination of Pesticide Residues in Produce. *J. AOAC Int.* **2019**, *86*, 412–431. [CrossRef]
11. Polgár, L.; Kmellár, B.; Garcia-Reyes, J.F.; Fodor, P. Comprehensive Evaluation of the Clean-up Step in the QuEChERS Procedure for the Multi-Residue Determination of Pesticides in Different Vegetable Oils Using LC-MS/MS. *Anal. Methods* **2012**, *4*, 1142–1148. [CrossRef]
12. Hou, R.; Jiao, W.; Xiao, Y.; Guo, J.; Lv, Y.; Tan, H.; Hu, J.; Wan, X. Novel use of PVPP in a modified QuEChERS extraction method for UPLC-MS/MS analysis of neonicotinoid insecticides in tea matrices. *Anal. Methods* **2015**, *7*, 5521–5529. [CrossRef]
13. Li, J.; Dong, F.; Xu, J.; Liu, X.; Li, Y.; Shan, W.; Zheng, Y. Enantioselective determination of triazole fungicide simeconazole in vegetables, fruits, and cereals using modified QuEChERS (quick, easy, cheap, effective, rugged and safe) coupled to gas chromatography/tandem mass spectrometry. *Anal. Chim. Acta* **2011**, *702*, 127–135. [CrossRef]
14. Li, L.; Li, W.; Qin, D.; Jiang, S.; Liu, F. Application of Graphitized Carbon Black to the QuEChERS Method for Pesticide Multiresidue Analysis in Spinach. *J. AOAC Int.* **2009**, *92*, 538–547. [CrossRef]
15. Rajski, Ł.; Lozano, A.; Uclés, A.; Ferrer, C.; Fernández-Alba, A.R. Determination of pesticide residues in high oil vegetal commodities by using various multi-residue methods and clean-ups followed by liquid chromatography tandem mass spectrometry. *J. Chromatogr. A* **2013**, *1304*, 109–120. [CrossRef]
16. Hou, X.; Han, M.; Dai, X.; Yang, X.; Yi, S. A multi-residue method for the determination of 124 pesticides in rice by modified QuEChERS extraction and gas chromatography–tandem mass spectrometry. *Food Chem.* **2013**, *138*, 1198–1205. [CrossRef]

17. Kim, L.; Baek, S.; Son, K.; Kim, E.; Noh, H.H.; Kim, D.; Oh, M.-s.; Moon, B.-c.; Ro, J.-H. Optimization of a Simplified and Effective Analytical Method of Pesticide Residues in Mealworms (*Tenebrio molitor* Larvae) Combined with GC–MS/MS and LC–MS/MS. *Molecules* **2020**, *25*, 3518. [CrossRef]
18. Lehotay, S.; Mastovska, K.; Lightfield, A. Use of Buffering Other Means to Improve Results of Problematic Pesticides in a Fast Easy Method for Residue Analysis of Fruits Vegetables. *J. AOAC Int.* **2005**, *88*, 615–629. [CrossRef]
19. Shin, Y.; Kim, C.; Baek, S.; Kim, L.; Son, K.-A.; Lee, H.-D.; Kim, D.; Kim, J.-H.; Noh, H. Liquid Chromatography-Tandem Mass Spectrometry for the Simultaneous Analysis of 353 Pesticides in the Edible Insect *Tenebrio molitor* Larvae (Mealworms). *Molecules* **2020**, *25*, 5866. [CrossRef]
20. Lehotay, S. QuEChERS Sample Preparation Approach for Mass Spectrometric Analysis of Pesticide Residues in Foods. *Methods Mol. Biol.* **2011**, *747*, 65–91. [CrossRef]
21. Li, J.; Sun, M.; Chang, Q.; Hu, X.; Kang, J.; Fan, C. Determination of Pesticide Residues in Teas via QuEChERS Combined with Dispersive Liquid–Liquid Microextraction Followed by Gas Chromatography–Tandem Mass Spectrometry. *Chromatographia* **2017**, *80*, 1447–1458. [CrossRef]
22. Wang, Y.; Miao, X.; Wei, H.; Liu, D.; Xia, G.; Yang, X. Dispersive Liquid-Liquid Microextraction Combined with Gas Chromatography-Mass Spectrometry for the Determination of Multiple Pesticides in Celery. *Food Anal. Methods* **2016**, *9*, 2133–2141. [CrossRef]
23. Łozowicka, B.; Rutkowska, E.; Jankowska, M. Influence of QuEChERS modifications on recovery and matrix effect during the multi-residue pesticide analysis in soil by GC/MS/MS and GC/ECD/NPD. *Env. Sci. Pollut. Res. Int.* **2017**, *24*, 7124–7138. [CrossRef]
24. European Commission. Analytical Quality Control and Method Validation Procedures for Pesticide Residues Analysis in Food and Feed (SANTE/11312/2021). Available online: https://www.eurl-pesticides.eu/userfiles/file/EurlALL/SANTE_11312_2021.pdf (accessed on 17 April 2025).
25. Marchi, I.; Viette, V.; Badoud, F.; Fathi, M.; Saugy, M.; Rudaz, S.; Veuthey, J.-L. Characterization and classification of matrix effects in biological samples analyses. *J. Chromatogr. A* **2010**, *1217*, 4071–4078. [CrossRef]
26. *EN 15662:2018*; Foods of Plant Origin—Determination of Pesticide Residues Using GC-MS and/or LC-MS/MS Following Acetonitrile Extraction/Partitioning and Clean-Up by Dispersive SPE—QuEChERS-Method. CEN: Brussels, Belgium, 2018. Available online: https://www.en-standard.eu/bs-en-15662-2018-foods-of-plant-origin-multimethod-for-the-determination-of-pesticide-residues-using-gc-and-lc-based-analysis-following-acetonitrile-extraction-partitioning-and-clean-up-by-dispersive-spe-modular-quechers-method/ (accessed on 17 April 2025).
27. Trufelli, H.; Palma, P.; Famiglini, G.; Cappiello, A. An overview of matrix effects in liquid chromatography-mass spectrometry. *Mass. Spectrom. Rev.* **2011**, *30*, 491–509. [CrossRef]
28. Chawla, S.; Patel, H.; Gor, N.; Vaghela, K.; Solanki, P.; Shah, P. Evaluation of Matrix Effects in Multiresidue Analysis of Pesticide Residues in Vegetables and Spices by LC-MS/MS. *J. AOAC Int.* **2017**, *100*, 616–623. [CrossRef]
29. Ferrer, C.; Lozano, A.; Agüera, A.; Girón, A.J.; Fernández-Alba, A.R. Overcoming matrix effects using the dilution approach in multiresidue methods for fruits and vegetables. *J. Chromatogr. A* **2011**, *1218*, 7634–7639. [CrossRef]
30. Mauer, L.; Forny, L.; Meunier, V.; Taylor, L. Optimizing the Quality of Food Powder Products: The Challenges of Moisture-Mediated Phase Transformations. *Annu. Rev. Food Sci. Technol.* **2019**, *10*, 457–478. [CrossRef]
31. Noh, H.H.; Kim, C.J.; Kim, S.-H.; Eun, H.-R.; Shin, Y.; Jeong, W.T. GC–MS/MS-based multiresidue pesticide analysis in mealworm (*Tenebrio molitor*) larvae: Optimization of standard QuEChERS-based method to minimize matrix effects. *Food Chem. X* **2025**, *27*, 102386. [CrossRef]

molecules

MDPI

Article

Determination of Phenylurea Herbicides in Water Samples by Magnet-Integrated Fabric Phase Sorptive Extraction Combined with High Performance Liquid Chromatography

Natalia Manousi [1], Apostolia Tsiasioti [1,2], Abuzar Kabir [3] and Erwin Rosenberg [1,*]

[1] Institute of Chemical Technologies and Analytics, TU Wien, 1060 Vienna, Austria; natalia.manousi@tuwien.ac.at (N.M.); tsiasioti@chem.auth.gr (A.T.)
[2] Laboratory of Analytical Chemistry, Department of Chemistry, Aristotle University of Thessaloniki, 54124 Thessaloniki, Greece
[3] Department of Chemistry and Biochemistry, International Forensic Research Institute, Florida International University, Miami, FL 33131, USA; akabir@fiu.edu
* Correspondence: egon.rosenberg@tuwien.ac.at

Abstract

In this study, a magnet-integrated fabric phase sorptive extraction (MI-FPSE) protocol was developed in combination with high pressure liquid chromatography—diode array detection (HPLC-DAD) for the simultaneous determination of five phenylurea pesticides (i.e., chlorbromuron, diuron, linuron, metoxuron, monuron) in environmental water samples. To produce the MI-FPSE device, two individual sol-gel coated carbowax 20 M (CW 20 M) cellulose membranes were fabricated and stitched to each other, while a magnetic rod was inserted between them to give the resulting device the ability to spin and serve as a stand-alone microextraction platform. The adsorption and desorption step of the MI-FPSE protocol was optimized to achieve high extraction efficiency and the MI-FPSE-HPLC-DAD method was validated in terms of linearity, sensitivity, selectivity, accuracy, and precision. The limits of detection (LODs) were found to be 0.3 μg L^{-1}. The relative recoveries were 85.2–110.0% for the intra-day and 87.7–103.2% for the inter-day study. The relative standard deviations were better than 13% in all cases. The green character and the practicality of the developed procedure were assessed using ComplexGAPI and Blue Analytical Grade Index metric tools, showing good method performance. Finally, the developed method was successfully used for the analysis of tap, river, and lake water samples.

Keywords: phenylurea herbicides; magnet-integrated fabric phase extraction; water; environmental analysis; HPLC-DAD

1. Introduction

Phenylurea herbicides (PUHs) are commonly used in agriculture to control weeds both before and after crop emergence in a variety of grain and vegetable crops, such as rice, potato, cotton, corn, and soybean [1,2]. The general formula of these pesticides typically includes a phenyl ring substituted with a urea functional group. These compounds also contain various substituents that determine their chemical properties and biological activity which greatly affect their action [3]. Despite their advantages in enhancing the production, the use of PUHs can have a negative impact since they can persist and accumulate in the environment for a long time [1,2]. At elevated concentrations, PUHs can be toxic to aquatic organisms, birds, and humans, posing significant risks to both ecosystems and public health [1–3]. For environmental and food safety, it is necessary to sensitively determine PUHs in environmental samples since pesticides are potential contaminants of surface freshwater [4,5].

For the determination of PUHs in real samples, chromatographic techniques such as high-performance liquid chromatography (HPLC) and gas chromatography (GC) combined with a wide variety of detection systems are typically employed. Among these, HPLC is commonly chosen since GC requires an additional derivatization step as PUHs are thermally labile compounds [2]. As pesticides occur at low concentration levels in real samples, sample preparation is typically required for their simultaneous extraction and preconcentration [6]. However, established sample preparation methodologies are tedious, time-consuming, and, more importantly, expend large quantities of resources that result in the generation of hazardous laboratory waste [7].

Currently, the adoption of greener approaches for pesticide extraction and sample preparation is a major focus of research [8]. In this context, different miniaturized extraction and microextraction techniques have emerged and are used for pesticide extraction. Microextraction is a green alternative to large-scale extraction techniques such as liquid-liquid extraction and solid-phase extraction, due to their miniaturization which results in reduced resources requirement, reusability, and enhanced user safety [9]. Typical examples of such sample preparation techniques include solid-phase microextraction [10], stir bar sorptive extraction [11], magnetic solid-phase extraction [12], dispersive liquid-liquid microextraction [9], and fabric phase sorptive extraction (FPSE) [13]. These techniques comply with many of the principles of Green Sample Preparation (GSP) that were proposed in 2022 by López-Lorente et al. [7]. These principles prioritize the selection of safer chemicals and materials that show renewability, recyclability, and reusability, the minimization of waste generation and energy demand, the miniaturization of sample preparation, the protection of the operator, and the achievement of high sample throughput among others.

Among these techniques, FPSE has attracted the interest of analytical chemists due to its beneficial characteristics. In this technique, a permeable and flexible fabric substrate is coated with a sol–gel organic–inorganic hybrid sorbent. This results in a microextraction device that can be directly immersed into the sample solution and extract the target analytes. A plethora of sorbents with different chemical characteristics (i.e., polar, non-polar, medium polar, ion exchangers etc.) has been developed to extract analytes with different chemical properties [14]. Magnet-integrated fabric phase sorptive extraction (MI-FPSE) is a modification of the conventional FPSE technique that integrates the microextraction and stirring mechanism into one unit. For this purpose, two FPSE membranes are sandwiched together, and a metallic magnetic stirrer is integrated into the device [15]. MI-FPSE techniques offer several notable advantages, including ease of handling, rapid and simultaneous preparation of multiple samples, environmental sustainability, and cost-effectiveness [16]. Both FPSE and MI-FPSE comply with many of the requirements of GSP.

In this work, MI-FPSE was used in combination with high-performance liquid chromatography (HPLC-DAD) for the determination of PUHs in environmental water samples. To the best of our knowledge, no previous protocol employing FPSE or MI-FPSE has been proposed in environmental analysis for the extraction of PUHs, which serves as a significant class of pesticides. For this purpose, different sol-gel coated MI-FPSE membranes were examined, and the sample preparation step was optimized. Accordingly, the MI-FPSE-HPLC-DAD was validated and applied for the analysis of river and lake water samples. The green character and the practicality of the proposed method were also examined using Complementary Green Analytical Procedure Index (ComplexGAPI) [17] and Blue Applicability Grade Index (BAGI) [18].

2. Results and Discussion

2.1. Optimization of the MI-FPSE Procedure

The main parameters that affect the extraction performance of PUHs from water samples using FPSE were investigated. The variables that were studied included: the sol-gel sorbent and its size, extraction time, sample volume, stirring rate, salt concentration, extraction solvent, volume of the eluent, and elution time. Each factor was independently studied, while the other variables were kept constant. The following conditions were used during method optimization: 1.5 cm × 1.5 cm^2 FPSE size, 30 min extraction time, 20 mL sample volume, 1000 rpm stirring rate, 0% *w/v* NaCl, 1 mL of MeOH extraction solvent, and 5 min elution time.

In order to find the optimum FPSE membrane for the extraction of PUHs from water samples three different sol-gel coated FPSE membranes were examined, with different properties. For this purpose, polar sol-gel carbowax 20 M (CW 20 M), medium polar sol-gel poly (tetrahydrofuran) (PTHF) and non-polar sol-gel octadecyl (C_{18}) membranes were evaluated. The experimental results are shown in Figure S1. As can be observed, the highest extraction recovery was obtained using the sol–gel CW 20 M FPSE membrane for all analytes. This can be attributed to the polar characteristics of PUHs due to the presence of the urea functional group. CW 20 M provides different intermolecular interactions including dipole–dipole interaction, hydrogen bonding, and London dispersion towards PUHs, favoring their extraction [19]. The utilization of a polar sol-gel material was expected to result in the highest extraction efficiency due to the possible existence of dipole-dipole interactions and hydrogen bonding forces. Thus, the sol–gel coated CW 20 M cellulose FPSE membrane was finally chosen for further investigation.

The effect of the size of the membranes was also studied, since the sorbent amount can affect the extraction recoveries of the target analytes [16]. For this purpose, four different dimensions were examined: 1.0 × 1.0, 1.5 × 1.5, 2.0 × 2.0, and 2.5 × 2.5 cm × cm, that correspond to 1.0, 2.25, 4.0, and 6.25 cm^2. As shown in Figure S2, the extraction recoveries were enhanced by increasing the size and thus the amount of sorbent. Further investigation was performed using two circular membranes, stitched to each other, with r = 1 cm that corresponds to 6.28 cm^2, to fabricate a MI-FPSE device that resulted in higher extraction recoveries. Therefore, the circular membrane of a sol-gel coated CW 20 M cellulose MI-FPSE membrane was finally chosen for the extraction of PUHs from water samples.

2.1.1. Optimization of the Adsorption Step

The main variables that affect the adsorption process of the analytes were investigated. The adsorption time was examined at six different values in the range of 10 to 60 min. The results are shown in Figure 1. Since FPSE is an equilibrium-based extraction technique, the optimum adsorption time corresponds to the time that the analytes need to reach the equilibrium between the sample and the sorbent [14]. As can be observed, all analytes reached the equilibrium at 30 min. Any further increase in the adsorption time did not improve the extraction efficiency. Thus, 30 min was selected as a compromise between the sensitivity and the highest sample throughput.

The volume of the sample was investigated, as it affects the extraction recoveries. For this purpose, volumes between 10 and 100 mL were studied, and the value of 10 mL resulted in the highest extraction recoveries, as shown in Figure S3. On the contrary, by increasing the sample volume the preconcentration factors are enhanced (2.5 to 10-fold). For this reason, a volume of 25 mL of sample was chosen as a compromise between the extraction recoveries, the sample consumption, and the preconcentration factors for the selected analytes.

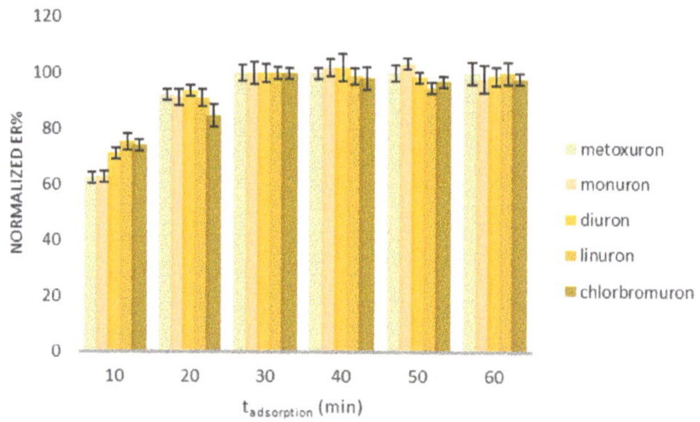

Figure 1. Evaluation of the adsorption time of the FPSE method. Concentration of the selected analytes: 100.0 μg L^{-1} (n = 3).

Another important factor investigated was the stirring rate, as it can promote the mass transfer of the analytes to the FPSE membrane, improving the adsorption and thus the extraction recoveries [20]. The stirring rate was examined at four different values in the range 0–1500 rpm. The experimental results are shown in Figure S4. As expected, the extraction recoveries were enhanced after increasing the stirring rate, with the highest values obtained at 1000 rpm. Further increase did not improve the mass transfer of the analytes. Therefore, 1000 rpm was selected as the optimal stirring rate.

Finally, the effect of salt addition was studied. For this reason, NaCl was added to the sample, as model electrolyte, in the range from 0% w/v to 30% w/v. As shown in Figure 2, the extraction recovery values were increased by increasing the salt concentration. This phenomenon is attributed to the salting out effect, in which the solubility of the analytes is reduced improving the analytes affinity to the sorptive phase [16]. Hence, 20% w/v NaCl was selected.

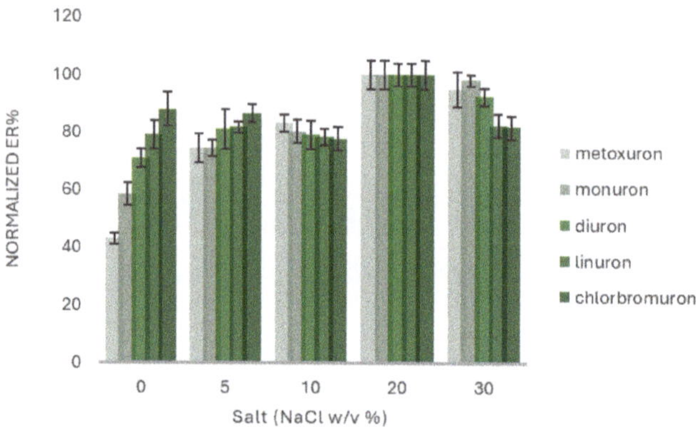

Figure 2. Evaluation of the effect of salt addition. Concentration of the selected analytes: 100.0 μg L^{-1} (n = 3).

2.1.2. Optimization of the Desorption Step

The main parameters that influence the desorption of the analytes from the FPSE membrane were investigated. The elution of the analytes was studied using four different eluents: methanol (MeOH), ethanol (EtOH), acetonitrile (ACN), and acetone (ACE). As shown in Figure S5, MeOH resulted in the most effective desorption of the analytes and was thus selected.

The volume of the eluent is an important factor because it influences the preconcentration factors. The volume of MeOH was investigated at four different values within the range 300–1500 μL. The amount of eluent has to be adequate in order to achieve sufficient desorption and as low as possible to minimize the waste of chemicals [20]. As shown in Figure S6, the extraction recoveries were stable. For this reason, an aliquot of 300 μL of MeOH was chosen, which is in accordance with the principles of green analytical chemistry (GAC).

Finally, optimization of the elution time was conducted in order to find the minimum time span required for sufficient desorption. The elution time was studied between 2 and 15 min and the experimental results are shown in Figure 3. The results were similar at all time spans investigated. Hence the minimum value of 2 min was selected as the optimum elution time to minimize sample preparation time.

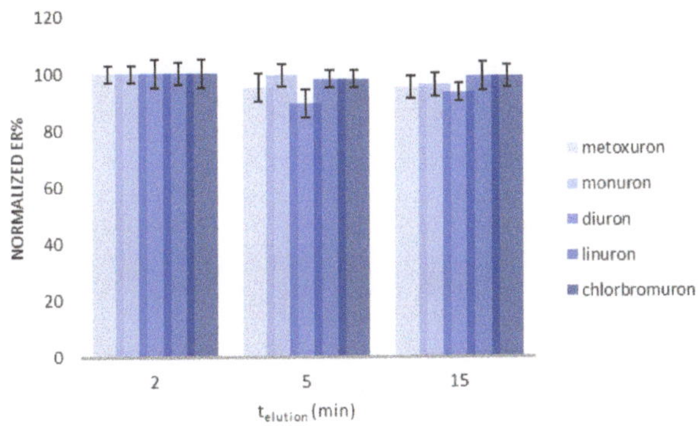

Figure 3. Evaluation of different desorption times. Concentration of the selected analytes: 100.0 μg L^{-1} (n = 3).

2.2. Method Validation

The developed MI-FPSE-HPLC-DAD method was validated in terms of linearity, accuracy, precision, limit of detection (LOD), and limit of quantification (LOQ). The analytical figures of merit are shown in Table 1. The linearity was evaluated by using a spiked water sample with the target analytes at concentration levels in the range of 1.0–100.0 μg L^{-1}. Least square linear regression analysis was applied to calculate the slopes, intercepts, and coefficients of determination (r^2) of PUHs. In all cases, the r^2 value is higher than 0.996, indicating a strong fit to the linear model and thus good linearity for the examined analytes.

Table 1. Analytical figures of merit of the proposed MI-FPSE-HPLC-DAD method.

Analyte	Regression Analysis	R^2	Working Range (μg L^{-1})	LOD (μg L^{-1})	LOQ (μg L^{-1})	ER% [1]	EF [2]
Metoxuron	y = 437x + 332	0.9960	1.0–100.0	0.3	1.0	60.3	50.1
Monuron	y = 694x − 92	0.9991	1.0–100.0	0.3	1.0	51.4	42.6
Diuron	y = 765x + 712	0.9968	1.0–100.0	0.3	1.0	61.4	50.9
Linuron	y = 728x + 438	0.9987	1.0–100.0	0.3	1.0	61.6	51.1
Chlorbromuron	y = 701x + 1046	0.9982	1.0–100.0	0.3	1.0	60.5	50.2

[1] ER%: extraction recovery % [2] EF: enhancement factor.

The LOQ values were defined as the lowest points of the calibration curves and corresponded to a signal-to-noise ratio of 10. The LOD values corresponded to a signal-to-noise ratio of 3 and were calculated by dividing the LOQ values by 3.3. Thus, the LOQs of PUHs were determined to be 1.0 μg L^{-1} and the LODs were 0.3 μg L^{-1}.

The enhancement factors (EF) were calculated by comparing the slope of the calibration curve before and after the preconcentration. The preconcentration factor (PF) was evaluated by dividing the initial volume (25 mL) and the final volume (300 μL) of the sample and was found to be 83. The extraction recoveries were calculated by comparing the EF of the method with the PF of each analyte. The ER% of the PUHs ranged between 51.4 and 61.6%, while their enhancement factors ranged between 42.6 and 51.1.

The precision and the accuracy of the proposed method were evaluated by using spiked water samples at two concentration levels at 5.0 and 50.0 μg L^{-1}. The experimental results are shown in Table 2. The intra-day precision was evaluated as the relative standard deviation (RSD%) of five repeated measurements and the inter-day precision by analysis of the spiked samples in three different days. The RSD values ranged from 2% to 7% for the intra-day precision and between 6% and 13% for the inter-day precision. The accuracy of the method was assessed by calculating the relative recovery (RR%), after the comparison of the found and added concentration of the spiked samples. The RR% was satisfactory and ranged from 85.2 to 110.0% for intra-day accuracy and between 87.7 and 103.2% for inter-day accuracy.

Table 2. Accuracy and precision of the proposed MI-FPSE-HPLC-DAD method.

Analyte	Added (μg L^{-1})	Intra-Day (n = 5)			Inter-Day (n = 3 × 3)		
		Found (μg L^{-1})	RSD%	RR%	Found (μg L^{-1})	RSD%	RR%
Metoxuron	5.0	4.6 ± 0.3	7	91.3	4.7 ± 0.3	7	93.1
	50.0	51.3 ± 1.2	2	105.5	50.5 ± 0.4	8	101.1
Monuron	5.0	4.7 ± 0.3	6	93.3	4.6 ± 0.4	8	92.6
	50.0	55.0 ± 2.0	4	110.0	51.6 ± 0.6	11	103.2
Diuron	5.0	4.8 ± 0.3	6	96.0	4.6 ± 0.3	7	92.8
	50.0	52.4 ± 1.8	3	104.8	49.1 ± 0.4	9	98.2
Linuron	5.0	4.3 ± 0.2	4	85.2	4.5 ± 0.3	7	89.1
	50.0	52.4 ± 0.9	2	104.8	48.4 ± 0.6	12	96.8
Chlorbromuron	5.0	4.3 ± 0.1	3	85.7	4.4 ± 0.3	6	87.7
	50.0	52.1 ± 1.0	2	104.1	47.8 ± 0.6	13	95.6

The performance of the sol-gel CW 20 M MI-FPSE membrane for the extraction of PUHs was examined as an indicator for the reusability of the membranes. Initially, potential carry-over was checked by using the membrane for the extraction of PUHs at the highest point of the calibration curve, followed by blank analysis. No unwanted carry-over phenomena were observed, indicating that the membrane could be reused. Accordingly, the membrane was repeatedly used for the extraction of PUHs from water at a concentration level of 50.0 μg L^{-1}. For the evaluation of the performance, the criterion of ±10% of the RR% was established. The results indicated that the MI-FPSE membrane can be used for at least 15 times, without losing significant performance.

2.3. *Investigation of Method's Green Character and Practicality*

The practicality of the developed procedure and its green character were examined with BAGI [18] and ComplexGAPI [17] metric tools, respectively. BAGI takes into consideration the main characteristics of an analytical method in terms of sample preparation and analytical determination, and it evaluates its applicability in terms of practicality. This tool generates an asteroid pictogram, and it produces a score that ranges from 25.0 (low practicality) to 100.0 (high practicality). An analytical method is considered practical if it results in a BAGI score higher than 60.0. The MI-FPSE-HPLC-DAD method produced a score of 70.0 (Figure 4a). The good practicality of the developed method can be attributed to the incorporation of different analytes in the analytical method, the simultaneous sample preparation of different samples, the high sample throughput, and the absence of additional steps in the sample preparation procedure. ComplexGAPI evaluates the green character of a method by evaluating the environmental impact of the sample pretreatment steps, the chemicals, instrumentation, sample preparation, and method type, while it can also assess the greenness of the preparation of the new extraction phase used in MI-FPSE. As shown in Figure 4b, the proposed method showed good compliance with greenness requirements due to the selection of a microextraction technique that reduces consumption of chemicals and waste generation.

(a) (b)

Figure 4. BAGI pictogram (**a**) and ComplexGAPI pictogram (**b**) of the proposed MI-FPSE-HPLC-DAD method.

2.4. Comparison with Other Studies

A comparison of the proposed MI-FPSE-HPLC-DAD method with other extraction methods developed for the determination of PUHs in water samples was conducted. As it can be observed in Table 3 the LOD of the proposed method is lower than the respective LOD in refs. [21–23], but higher than the methods in refs. [11,24–26]. However, it should be noted that the later methods contain an additional evaporation step that enhanced the preconcentration but also the complexity of sample preparation, reducing their practicality. This is clearly reflected in their BAGI scores which ranged between 50.0 and 67.5, which were lower than the value obtained for this study (i.e., 70.0). Due to the evaporation step, ref. [25] shows higher EFs contrary to the developed method, in combination with the high volume of sample applied. Similar EF values are reported for methods in refs. [11,24]. Regarding the accuracy, the present study shows better recoveries in comparison to the most of the works (refs. [11,22–26]). In addition, the herein developed method exhibited comparable precision with the other methods. Finally, the majority of the other methods studied the most common PUHs, mainly monuron and linuron, while chlorbromuron was only included in the present study.

Table 3. Comparison of the proposed MI-FPSE-HPLC-DAD method and with other HPLC methods for the determination of PUHs in water.

Sample Preparation	PUHs	Analytical Technique	Sample Volume (mL)	RR%	RSD%	EF	LOD (μg L^{-1})	BAGI Pictogram	Ref.
Ionic liquid based on dispersive liquid-phase microextraction	Linuron, diuron	HPLC-DAD	10	93.2–98.6	8.68–10.78	N.A. [1]	0.66–0.74	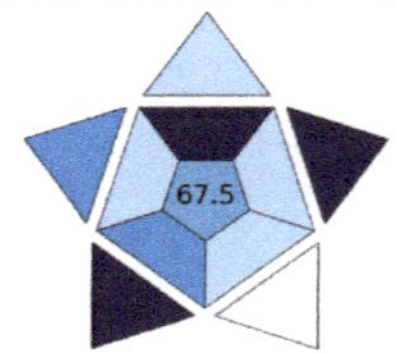	[21]
Supramolecular solvent microextraction	Monuron, linuron, isoproturon	HPLC-UV	20	80.0–99.0	<7.2 (intra) <6.5 (inter)	N.A.	10–30		[22]
Automated magnetic micro solid-phase extraction	Monuron, linuron, tebuthiuron, isoproturon,	HPLC-UV	20	80–116	<3 (intra) <6 (inter)	40–57	0.04	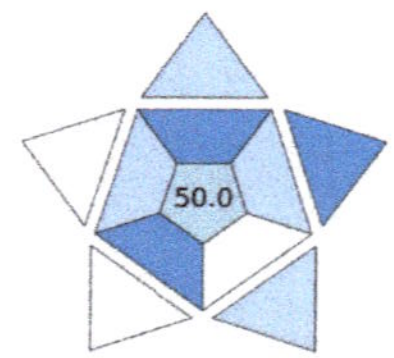	[24]
Functionalized coated stir bar sorptive extraction	Chlortoluron, isoproturon, diuron, linuron	HPLC-DAD	10	80.0–114.8	<4.34 (intra) <8.57 (inter)	46–49	0.025–0.070		[11]

Table 3. *Cont.*

Sample Preparation	PUHs	Analytical Technique	Sample Volume (mL)	RR%	RSD%	EF	LOD (μg L^{-1})	BAGI Pictogram	Ref.
Magnetic micro solid-phase extraction	Monuron, chlorotoluron, isoproturon, monolinuron, buturon	HPLC-DAD	50	80.6–118.0	<9.6 (intra) <9.6 (inter)	126–227	0.0025–0.015		[25]
Magnetic micro solid-phase extraction	Monuron, chlortoluron, monolinuron, isoproturon, linuron	HPLC-UV	10	74.5–111.4	<12.7 (intra) <13.3 (inter)	N.A.	0.06–0.10		[26]
Solid phase extraction	Monuron, diuron, linuron, metoxuron, metazachlor	HPLC-UV	20	75.0–90.1	N.A.	N.A.	0.82–1.29		[23]
MI-FPSE	Metoxuron, monuron, diuron, linuron, chlorbromuron	HPLC-DAD	25	85.2–110.0	<7 (intra) <13% (inter)	51.4–61.6	0.3	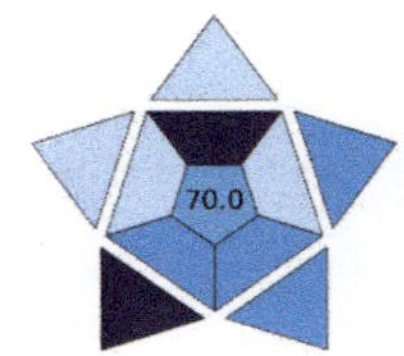	This study

[1] N.A.: not available.

2.5. Real Sample Analysis

To demonstrate the practical applicability of the newly developed MI-FPSE-HPLC-DAD method, analyses of various water samples were carried out. Table 4 summarizes the results for real sample analysis. Spiked water samples (50.0 μg L^{-1}) were also prepared and analyzed, while the RR% for each analyte was calculated. Since the RR% values were within the range of 86.5–114.3%, the proposed protocol can be used for the reliable determination of PUHs in different surface water samples.

Table 4. Determination of PUHs in water samples through the proposed method (average value ± standard deviation).

Analyte	Added (μg L^{-1})	Tap Water		River Water		Lake Water	
		Found (μg L^{-1})	RR%	Found (μg L^{-1})	RR%	Found (μg L^{-1})	RR%
Metoxuron	-	<LOD	-	<LOD	-	<LOD	-
	50.0	56.0 ± 1.1	112.0	48.4 ± 2.4	96.8	48.0 ± 1.4	96.1
Monuron	-	<LOD	-	<LOD	-	<LOD	-
	50.0	57.1 ± 0.4	114.3	46.3 ± 1.6	92.7	47.8 ± 0.3	95.6
Diuron	-	<LOD	-	<LOD	-	<LOD	-
	50.0	49.7 ± 3.4	99.5	53.6 ± 3.4	107.2	47.1 ± 3.3	94.1
Linuron	-	<LOD	-	<LOD	-	<LOD	-
	50.0	45.3 ± 1.6	90.6	50.1 ± 2.8	100.2	47.5 ± 3.4	94.9
Chlorbromuron	-	<LOD	-	<LOD	-	<LOD	-
	50.0	43.1 ± 2.9	86.5	46.1 ± 1.4	92.2	51.2 ± 3.0	102.4

3. Materials and Methods

3.1. Standards, Chemicals, and Samples

Chlorbromuron, diuron, linuron, metoxuron, monuron ($\geq$ 98.0%) were obtained from Supelco (Bellefonte, PA, USA). Figure S7 shows the chemical structures of the target analytes. For all analytes, a stock solution was prepared in MeOH at a concentration of 1000 mg L^{-1}. Working standard solutions were daily prepared in MeOH at appropriate concentration levels.

Methanol (MeOH), acetonitrile (ACN), acetone (ACE), ethanol (EtOH) and water of LC-MS grade were purchased from Honeywell (Charlotte, North Carolina, USA). Analytical grade sodium chloride was purchased from Merck Life Science (Darmstadt, Germany).

Environmental water samples (i.e., lake, river, and tap water) were collected in the municipal area of Vienna, Austria. The collection of the samples was conducted in amber-glass vials with no headspace. Sample containers were stored at 4 °C. No filtration of the samples was necessary prior to their analysis.

For the fabrication of the MI-FPSE device, two sol-gel coated CW 20 M MI-FPSE membranes were sandwiched together, and a metallic magnetic stirrer was integrated between them. The synthesis of the sol-gel CW 20 M sorbent has been described elsewhere [27].

3.2. HPLC-DAD Analysis

An HPLC–DAD system from Shimadzu (Kyoto, Japan) was employed for the quantification of the target analytes. For this purpose, a column oven (CTO-20AC), two pumps (LC-20ADP), a communication bus module (CBM-20A), a degassing unit (DGU-20A3), a diode array detector (SPD-M20A), and an autosampler (SIL-20A) were used. LCMS solution version 3.81 software was used for system handling, data acquisition, and data processing. The separation of the PUHs was performed using a Kinetex C_{18} 100A (2.6 μm, 30 × 4.6 mm) column (Phenomenex, CA, USA). Water (A) and ACN (B) were used as mobile phase. The initial mobile phase composition was 75:25 (A:B, v/v) and it was maintained for 2.5 min. Then, the composition changed to 55:45 (A:B, v/v) until 5.0 min and it was kept constant up to 6.0 min. At 6.0 min, the composition was changed until 7.0 min at 80:20 (A:B, *v*/*v*). The system returned to the initial parameters at 8.0 min and it was equilibrated until 12.0 min. The column was thermostated at 30 °C and the analytes were monitored at 245 nm. Using these parameters, the retention times for the analytes were as such: metoxuron 1.4 min, monuron 1.7 min, diuron 3.9 min, linuron 5.2 min, and chlorbromuron 5.4 min. Figure 5 shows a representative chromatogram of the separation of the PUHs. The system suitability parameters are shown in Table S1.

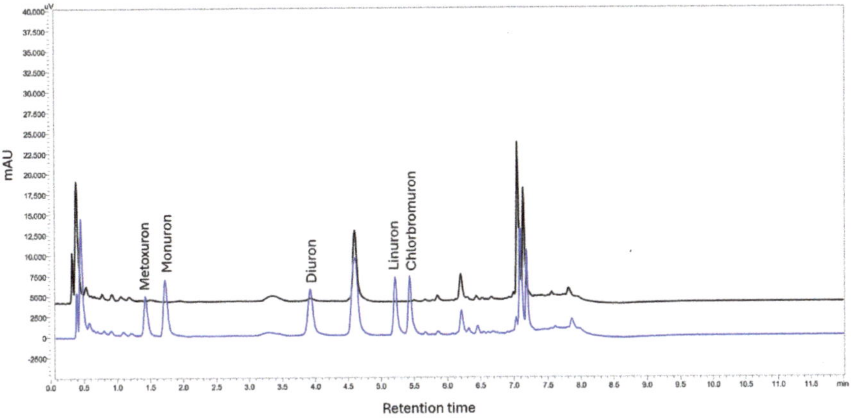

Figure 5. A representative chromatogram of a blank (black) and a spiked (c = 50.0 μg L^{-1}) river water sample (blue) analyzed through the proposed MI-FPSE-HPLC-DAD method. Retention times: metoxuron 1.4 min, monuron 1.7 min, diuron 3.9 min, linuron 5.2 min, and chlorbromuron 5.4 min.

3.3. MI-FPSE Procedure

Initially, the MI-FPSE membranes were activated through sequential immersion in 1 mL of an ACN/MeOH (50:50, *v*/*v*) mixture for 5 min and 1 mL of H_2O for 5 min. For the extraction of the target analytes, the activated membranes were immersed into 25 mL of sample solution (adjusted to 20% *w*/*v* NaCl) that was placed into a 40 mL glass vial. The adsorption took place within 30 min under stirring at a rate of 1000 rpm. Then, the membrane was removed, and the supernatant solution was discarded. The extraction medium was rinsed with water and dried with lint-free tissue paper. The target analytes were eluted by immersing the MI-FPSE membranes in an Eppendorf tube in 300 μL of methanol for 2 min. Following this timespan, the eluate was analyzed by HPLC-DAD. The MI-FPSE membranes were washed with 1 mL of ACN/MeOH (50:50, *v*/*v*), dried, and stored until their next use.

4. Conclusions

In this work, a simple and efficient MI-FPSE protocol was proposed for the extraction and preconcentration of five PUHs. The MI-FPSE procedure was combined with HPLC-DAD for water sample analysis. Sol-gel CW 20 M MI-FPSE media showed the highest extraction capability towards the target analytes. No additional pre-treatment (e.g., suspension treatment, centrifugation, filtration) or post-treatment (dilution, evaporation, reconstitution) was required, reducing the steps of analysis, and making the procedure less error prone. The developed method exhibited good performance characteristics in terms of sensitivity, accuracy, and precision. Moreover, the overall scheme exhibited handling simplicity, practicality, green character, and the ability to perform simultaneous sample preparation of different samples. Finally, the MI-FPSE membranes were reusable complying with the requirements of GAC and GSP. The proposed approach could be employed for the extraction of PUHs from other aqueous sample matrices (e.g., fruit juice, tea, herbal infusion samples etc.). However, in this case, the matrix-effect should be assessed, and matrix-matched calibration should be constructed. It is also worth-mentioning that the configuration of MI-FPSE favors the utilization of different FPSE membranes, sandwiched together, for the simultaneous extraction of analytes with a wide range of polarity. All things considered, MI-FPSE can serve as a powerful tool towards greener pesticide determination in environmental and food samples.

Supplementary Materials: The following supporting information can be downloaded at: https://www.mdpi.com/article/10.3390/molecules30153135/s1, Figure S1: Study of different sol-gel membranes. C18: octadecyl, PTHF: poly (tetrahydrofuran), CW 20 M: Carbowax 20, Figure S2: Evaluation of the size of the membrane. Concentration of the selected analytes: 100.0 μg L^{-1}, Figure S3: Study of the stirring rate of the proposed MI-FPSE method. Concentration of the selected analytes: 100.0 μg L^{-1}, Figure S4: Evaluation of elution solvent of the adsorbed analytes. Concentration of the selected analytes: 100.0 μg L^{-1}, Figure S5: Study of different sample volumes. Concentration of the selected analytes: 100.0 μg L^{-1}, Figure S6: Chemical structures of the selected analytes, Figure S7: Chemical structures of the selected analytes, Table S1: The system suitability parameters of the proposed MI-FPSE-HPLC-DAD method.

Author Contributions: Conceptualization, N.M., A.K. and E.R.; methodology, N.M., A.T., A.K. and E.R.; software, N.M. and A.T.; validation, N.M. and A.T.; formal analysis, N.M. and A.T.; investigation, N.M. and A.T.; resources, A.K. and E.R.; data curation, N.M. and A.T.; writing—original draft preparation, N.M. and A.T.; writing—review and editing, A.K. and E.R.; visualization, A.T.; supervision, A.K. and E.R.; project administration, E.R. All authors have read and agreed to the published version of the manuscript.

Funding: This research received no external funding.

Institutional Review Board Statement: Not applicable.

Informed Consent Statement: Not applicable.

Data Availability Statement: Data will be made available upon request.

Conflicts of Interest: The authors declare no conflicts of interest.

Abbreviations

The following abbreviations are used in this manuscript:

ACE	Acetone
ACN	Acetonitrile
BAGI	Blue Applicability Grade Index
C_{18}	Octadecyl
ComplexGAPI	Complementary Green Analytical Procedure Index
CW 20 M	Carbowax 20 M
EF	Enhancement factor
EtOH	Ethanol
FPSE	Fabric phase sorptive extraction
GAC	Green Analytical Chemistry
GC	Gas chromatography
GSP	Green Sample Preparation
HPLC	High performance liquid chromatography
HPLC-DAD	High performance liquid chromatography-diode array detection
LOD	Limit of Detection
LOQ	Limit of Quantification
MeOH	Methanol

MI-FPSE	Magnet integrated-fabric phase sorptive extraction
PF	Preconcentration factor
PTHF	Poly(tetrahydrofuran)
PUH	Phenylurea herbicide
RR	Relative recovery
RSD	Relative standard deviation

References

1. Mou, R.X.; Chen, M.X.; Zhi, J.L. Simultaneous determination of 15 phenylurea herbicides in rice and corn using HPLC with fluorescence detection combined with UV decomposition and post-column derivatization. *J. Chromatogr. B Anal. Technol. Biomed. Life Sci.* **2008**, *875*, 437–443. [CrossRef]
2. Li, J.; Xu, X.; Zhang, F.; Guo, W.; Wang, X.; Xie, Y.; Zhang, F. Urea-based magnetic porous organic frameworks as novel adsorbent for the enrichment of phenylurea herbicides in foods. *Food Chem.* **2023**, *425*, 136436. [CrossRef]
3. Mohanty, S.S.; Singh, P.; Nistala, S.; Mohanty, K. Understanding the environmental fate and removal strategies of phenylurea herbicides: A comprehensive review. *J. Hazard. Mater. Adv.* **2024**, *16*, 100496. [CrossRef]
4. Guo, L.; Hao, L.; Gao, T.; Wang, C.; Wu, Q.; Wang, Z. p-Phenylenediamine-modified graphene oxide as a sorbent for solid-phase extraction of phenylurea herbicides, nitroimidazoles, chlorophenols, phenylurea insecticides and phthalates. *Microchim. Acta* **2019**, *186*, 464. [CrossRef]
5. de Araújo, E.P.; Caldas, E.D.; Oliveira-Filho, E.C. Pesticides in surface freshwater: A critical review. *Environ. Monit. Assess.* **2022**, *194*, 452. [CrossRef]
6. Musarurwa, H.; Chimuka, L.; Tavengwa, N.T. Green pre-concentration techniques during pesticide analysis in food samples. *J. Environ. Sci. Health—Part B Pestic. Food Contam. Agric. Wastes* **2019**, *54*, 770–780. [CrossRef]
7. López-Lorente, Á.I.; Pena-Pereira, F.; Pedersen-Bjergaard, S.; Zuin, V.G.; Ozkan, S.A.; Psillakis, E. The ten principles of green sample preparation. *TrAC—Trends Anal. Chem.* **2022**, *148*, 116530. [CrossRef]
8. Nasiri, M.; Ahmadzadeh, H.; Amiri, A. Sample preparation and extraction methods for pesticides in aquatic environments: A review. *TrAC—Trends Anal. Chem.* **2020**, *123*, 115772. [CrossRef]
9. Ahmad, W.; Al-Sibaai, A.A.; Bashammakh, A.S.; Alwael, H.; El-Shahawi, M.S. Recent advances in dispersive liquid-liquid microextraction for pesticide analysis. *TrAC—Trends Anal. Chem.* **2015**, *72*, 181–192. [CrossRef]
10. Beltran, J.; López, F.J.; Hernández, F. Solid-phase microextraction in pesticide residue analysis. *J. Chromatogr. A* **2000**, *885*, 389–404. [CrossRef]
11. Han, J.H.; Cui, Y.Y.; He, X.Q.; Zhang, Y.; Yang, C.X. Fabrication of carboxyl functionalized microporous organic network coated stir bar for efficient extraction and analysis of phenylurea herbicides in food and water samples. *J. Chromatogr. A* **2021**, *1640*, 461947. [CrossRef]
12. Musarurwa, H.; Chimuka, L.; Tavengwa, N.T. Metal organic framework-based magnetic solid phase extraction of pesticides in complex matrices. *Microchem. J.* **2021**, *171*, 106907. [CrossRef]
13. Kaur, R.; Kaur, R.; Rani, S.; Malik, A.K.; Kabir, A.; Furton, K.G.; Samanidou, V.F. Rapid monitoring of organochlorine pesticide residues in various fruit juices and water samples using fabric phase sorptive extraction and gas chromatography-mass spectrometry. *Molecules* **2019**, *24*, 1013. [CrossRef]
14. Kabir, A.; Samanidou, V. Fabric Phase Sorptive Extraction: A Paradigm Shift Approach in Analytical and Bioanalytical Sample Preparation. *Molecules* **2021**, *26*, 865. [CrossRef]
15. Samanidou, V.F.; Kabir, A. Magnet Integrated Fabric Phase Sorptive Extraction (MI-FPSE): A Powerful Green(er) Alternative for Sample Preparation. *Analytica* **2022**, *3*, 439–447. [CrossRef]
16. Manousi, N.; Alampanos, V.; Ferracane, A.; Efstratiadis, G.; Kabir, A.; Furton, K.G.; Tranchida, P.Q.; Zachariadis, G.A.; Mondello, L.; Rosenberg, E.; et al. Magnet integrated fabric phase sorptive extraction as a stand-alone extraction device for the monitoring of benzoyl urea insecticides in water samples by HPLC-DAD. *J. Chromatogr. A* **2022**, *1672*, 463026. [CrossRef]
17. Płotka-Wasylka, J.; Wojnowski, W. Complementary green analytical procedure index (ComplexGAPI) and software. *Green Chem.* **2021**, *23*, 8657–8665. [CrossRef]
18. Manousi, N.; Wojnowski, W.; Płotka-Wasylka, J.; Samanidou, V.F. Blue applicability grade index (BAGI) and software: A new tool for the evaluation of method practicality. *Green Chem.* **2023**, *25*, 7598–7604. [CrossRef]
19. Jain, B.; Jain, R.; Kabir, A.; Sharma, S. Rapid Determination of Non-Steroidal Anti-Inflammatory Drugs in Urine Samples after In-Matrix Derivatization and Fabric Phase Sorptive Extraction-Gas Chromatography-Mass Spectrometry Analysis. *Molecules* **2022**, *27*, 7188. [CrossRef]
20. Ulusoy, H.İ.; Köseoğlu, K.; Kabir, A.; Ulusoy, S.; Locatelli, M. Fabric phase sorptive extraction followed by HPLC-PDA detection for the monitoring of pirimicarb and fenitrothion pesticide residues. *Microchim. Acta* **2020**, *187*, 337. [CrossRef]

21. Wang, S.; Ren, L.; Liu, C.; Ge, J.; Liu, F. Determination of five polar herbicides in water samples by ionic liquid dispersive liquid-phase microextraction. *Anal. Bioanal. Chem.* **2010**, *397*, 3089–3095. [CrossRef]
22. Amir, S.; Shah, J.; Jan, M.R. Supramolecular solvent microextraction of phenylurea herbicides from environmental samples. *Desalin. Water Treat.* **2019**, *148*, 202–212. [CrossRef]
23. Singh, B.; Kaur, M.; Malik, A.K. Determination of Phenylurea Herbicides in Tap Water and Soft Drink Samples by HPLC–UV and Solid-Phase Extraction. *LCGC Eur.* **2012**, *23*, 120–129.
24. Saraji, M.; Mohammadipour, L.; Mehrafza, N. An effective configuration for automated magnetic micro solid-phase extraction of phenylurea herbicides from water samples followed by high-performance liquid chromatography. *J. Chromatogr. A* **2020**, *1617*, 460829. [CrossRef]
25. Hong, K.; Huang, Y.; Zheng, L.; Zheng, X.; Huang, X. One-pot fabrication of poly (ionic liquid)s functionalized magnetic adsorbent for efficient enrichment of phenylurea herbicides in environmental waters. *Anal. Chim. Acta* **2022**, *1198*, 339549. [CrossRef]
26. Zhang, X.; Liu, J.; Zhang, H.; Zhang, Q.; Shen, J.; Wei, Y.; Wang, C. Facile construction of a stable core-shell spherically magnetic polyimide covalent organic framework for efficient extraction of phenylurea herbicides. *Talanta* **2024**, *275*, 126184. [CrossRef]
27. Tartaglia, A.; Locatelli, M.; Kabir, A.; Furton, K.G.; Macerola, D.; Sperandio, E.; Piccolantonio, S.; Ulusoy, H.I.; Maroni, F.; Bruni, P.; et al. Comparison between Exhaustive and Equilibrium Extraction Using Different SPE Sorbents and Sol-Gel Carbowax 20M Coated FPSE Media. *Molecules* **2019**, *24*, 382. [CrossRef]

MDPI AG
Grosspeteranlage 5
4052 Basel
Switzerland
Tel.: +41 61 683 77 34

Molecules Editorial Office
E-mail: molecules@mdpi.com
www.mdpi.com/journal/molecules

www.ingramcontent.com/pod-product-compliance
Lightning Source LLC
Chambersburg PA
CBHW062059220926
51850CB00010B/2827
* 9 7 8 3 7 2 5 8 7 0 0 4 2 *